David Hunger

A Bose-Einstein condensate coupled to a micromechanical oscillator

David Hunger

A Bose-Einstein condensate coupled to a micromechanical oscillator

Exploring novel interfaces for ultracold atoms

Südwestdeutscher Verlag für Hochschulschriften

Impressum/Imprint (nur für Deutschland/ only for Germany)
Bibliografische Information der Deutschen Nationalbibliothek: Die Deutsche Nationalbibliothek verzeichnet diese Publikation in der Deutschen Nationalbibliografie; detaillierte bibliografische Daten sind im Internet über http://dnb.d-nb.de abrufbar.

Alle in diesem Buch genannten Marken und Produktnamen unterliegen warenzeichen-, marken- oder patentrechtlichem Schutz bzw. sind Warenzeichen oder eingetragene Warenzeichen der jeweiligen Inhaber. Die Wiedergabe von Marken, Produktnamen, Gebrauchsnamen, Handelsnamen, Warenbezeichnungen u.s.w. in diesem Werk berechtigt auch ohne besondere Kennzeichnung nicht zu der Annahme, dass solche Namen im Sinne der Warenzeichen- und Markenschutzgesetzgebung als frei zu betrachten wären und daher von jedermann benutzt werden dürften.

Verlag: Südwestdeutscher Verlag für Hochschulschriften Aktiengesellschaft & Co. KG
Dudweiler Landstr. 99, 66123 Saarbrücken, Deutschland
Telefon +49 681 37 20 271-1, Telefax +49 681 37 20 271-0
Email: info@svh-verlag.de
Zugl.: München, LMU, Diss., 2010

Herstellung in Deutschland:
Schaltungsdienst Lange o.H.G., Berlin
Books on Demand GmbH, Norderstedt
Reha GmbH, Saarbrücken
Amazon Distribution GmbH, Leipzig
ISBN: 978-3-8381-1836-9

Imprint (only for USA, GB)
Bibliographic information published by the Deutsche Nationalbibliothek: The Deutsche Nationalbibliothek lists this publication in the Deutsche Nationalbibliografie; detailed bibliographic data are available in the Internet at http://dnb.d-nb.de.

Any brand names and product names mentioned in this book are subject to trademark, brand or patent protection and are trademarks or registered trademarks of their respective holders. The use of brand names, product names, common names, trade names, product descriptions etc. even without a particular marking in this works is in no way to be construed to mean that such names may be regarded as unrestricted in respect of trademark and brand protection legislation and could thus be used by anyone.

Publisher: Südwestdeutscher Verlag für Hochschulschriften Aktiengesellschaft & Co. KG
Dudweiler Landstr. 99, 66123 Saarbrücken, Germany
Phone +49 681 37 20 271-1, Fax +49 681 37 20 271-0
Email: info@svh-verlag.de

Printed in the U.S.A.
Printed in the U.K. by (see last page)
ISBN: 978-3-8381-1836-9

Copyright © 2010 by the author and Südwestdeutscher Verlag für Hochschulschriften Aktiengesellschaft & Co. KG and licensors
All rights reserved. Saarbrücken 2010

meiner Familie

Zusammenfassung

In dieser Dissertation wird die Kopplung zwischen ultrakalten Atomen und mikromechanischen Oszillatoren untersucht. In unserem Experiment positionieren wir ein Bose-Einstein Kondensat (BEC) mithilfe Chip-basierter magnetischer Mikrofallen nahe der Oberfläche eines mikromechanischen Balkenresonators. Wir zeigen, dass attraktive Oberflächenkräfte dazu verwendet werden können, Biegeschwingungen des Resonators an kollektive Schwingungen der Atome in der Falle zu koppeln.

Die Kopplung ermöglicht resonante Anregung mehrerer spektral getrennter mechanischer Moden des BECs. Wir beobachten unter anderem die Dipolmode und eine Kompressionsmode. Als Signatur für die Anregung atomarer Bewegung dienen erhöhte Fallenverluste, womit Schwingungen des Resonators bis zu minimalen Amplituden von 13 nm detektiert werden können. Unter Verwendung eines Schemas zur direkten Auslese der atomaren Bewegung sollte die Sensitivität um etwa zwei Grössenordnungen verbessert werden können.

Zur Charakterisierung der Wechselwirkung zwischen Atomen und Resonator entwickeln wir eine Methode, mit der die absolute Stärke der Oberflächenkräfte bestimmt werden kann. Dazu analysieren wir Messungen von Fallenverlusten an der Resonatoroberfläche und Messungen zu resonanter BEC Anregung auf beiden Seiten des Resonators. Damit lässt sich die Kopplungskonstante der Wechselwirkung quantitativ bestimmen. Die Messungen werden in Abständen von $0.8 - 2.5$ μm von der Oberfläche durchgeführt, so dass äußerste Positioniergenauigkeit erforderlich ist. Wir erreichen eine Reproduzierbarkeit der Fallenposition von unter 6 nm, deutlich kleiner als typische BEC Durchmesser von 500 nm. Präzise Positionierung ist grundlegend für eine Vielzahl weiterer Experimente, wie z.B. Messungen lokaler Atom-Oberflächenkräfte, rastermikroskopische Oberflächenabbildung mit Atomen und Kopplung von Atomen an Festkörpersysteme über elektromagnetische Nahfelder.

Gekoppelte Systeme aus ultrakalten Atomen und kryogenen mechanischen Oszillatoren werden als vielversprechendes hybrides Quantensystem diskutiert. Wir zeigen verschiedene Systeme auf, die eine Kopplung im Quantenregime ermöglichen können. Zur Kopplung über Oberflächenkräfte bedarf es keiner Spiegel, Magnete oder Elektroden auf dem Oszillator, sodass Atome an molekulare Resonatoren wie Kohlenstoffnanoröhrchen gekoppelt werden könnten. Alternativ untersuchen wir die magnetische Kopplung eines Nanoresonators an den Spin eines BECs. In diesen Systemen können die Atome den Oszillator signifikant beeinflussen, wodurch Kontrolle und Manipulation des Oszillators möglich wird. Darüberhinaus zeigen wir auf, dass das Regime starker Kopplung erreicht werden kann.

Abstract

This thesis reports experiments on the interaction between a Bose-Einstein condensate (BEC) of magnetically trapped ^{87}Rb atoms and the motion of a micromechanical oscillator. We make use of the exceptional control provided by chip-based magnetic microtraps to approach a microcantilever with a BEC to about one micrometer distance, where atom-surface forces play an important role. We show both theoretically and experimentally that the attractive forces close to the oscillator's surface can be used to coherently couple mechanical motion of the cantilever to collective motion of the atoms in the trap.

We observe resonant coupling to several well-resolved mechanical modes of the condensate, including in particular the center of mass mode and the breathing mode. We use trap loss as the simplest way to detect the atomic motion induced by the coupling. With this method we are able to sense cantilever oscillations with a minimum resolvable amplitude of 13 nm. We investigate the effects that limit such coupling experiments and find a quantitative explanation for our observations. We propose that the sensitivity could be improved by about two orders of magnitude by using an improved readout scheme for the atoms.

To quantify the atom-cantilever interaction we develop a method to characterize the absolute strength of the surface forces. We analyze measurements of atom loss in the static surface potential and loss induced by cantilever motion on both sides of the cantilever. This allows us to infer the value of the coupling constant that describes the interaction. The measurements are performed at a distance of $0.8 - 2.5$ μm from the surface, which requires exceptional precision in the positioning of the atoms. We achieve a positioning reproducibility below 6 nm, much less than the typical diameter of 500 nm of the condensates. Such high control is an important prerequisite also for other experiments such as measurements of local atom-surface forces, scanning surface microscopy with atoms, and the coupling of atoms to solid state systems through local electromagnetic fields.

Atoms coherently coupled to cryogenic mechanical oscillators are considered as promising hybrid quantum systems. We discuss different schemes that could enable atom-cantilever coupling at the quantum level. Coupling via surface forces does not require mirrors, electrodes, or magnets on the oscillator and could thus be employed to couple atoms to molecular-scale oscillators such as carbon nanotubes. Alternatively, we discuss the magnetic coupling of a nanomechanical resonator to the spin of a BEC. In both settings, back action of the atoms on the mechanical oscillator can become significant, enabling manipulation and control of the oscillator. We furthermore investigate the conditions required to achieve the strong coupling limit.

Contents

1. **Introduction**   1

2. **Atoms in magnetic chip traps**   7
   - 2.1. Magnetic microtraps 7
     - 2.1.1. Magnetic trapping of neutral atoms 8
     - 2.1.2. Trap configurations 10
   - 2.2. Properties of Bose-Einstein condensates 15
     - 2.2.1. Gross-Pitaevskii description 17
     - 2.2.2. Condensate excitations 18
     - 2.2.3. Condensate motion in anharmonic traps 24
   - 2.3. Surface forces 26
     - 2.3.1. Van der Waals-London and Casimir-Polder Potential 26
     - 2.3.2. Adsorbates and stray charges 29
   - 2.4. Effects of surface forces on atom trapping 35
     - 2.4.1. Potential deformation 35
     - 2.4.2. Sudden loss and surface evaporation 37
     - 2.4.3. Tunneling 38
     - 2.4.4. Quantum reflection 39
   - 2.5. Atom loss and heating 40
     - 2.5.1. Collisional loss 40
     - 2.5.2. Thermal magnetic near-field noise 41
     - 2.5.3. Technical heating 44

3. **Micro- and nanomechanical oscillators**   47
   - 3.1. Fundamental properties 47
     - 3.1.1. Modefunction, resonance frequency, and effective mass . . . 48
     - 3.1.2. Thermal motion 50
     - 3.1.3. Dissipation 51
   - 3.2. Detection and manipulation of mechanical motion 53
   - 3.3. Quantum states of mechanical oscillators 55
     - 3.3.1. Decoherence 57
     - 3.3.2. The size of a superposition 58

4. Setup and BEC production — 61
- 4.1. Atom chip with microcantilever — 61
  - 4.1.1. The atom chip — 62
  - 4.1.2. Resonator characterization and readout — 66
- 4.2. Vacuum, Lasersystem, Electronics — 70
  - 4.2.1. Vacuum setup — 70
  - 4.2.2. Laser system — 72
  - 4.2.3. Current sources and magnetic field coils — 75
- 4.3. Experimental sequence for BEC preparation — 76
- 4.4. Atoms close to the surface — 84
  - 4.4.1. Trap lifetime vs. atom-surface distance — 84
  - 4.4.2. BEC lifetime vs. trap frequency — 85
  - 4.4.3. Trap frequency measurements and trap simulation — 87

5. BEC-resonator coupling via surface forces — 89
- 5.1. Coupling via surface forces — 89
  - 5.1.1. Excitation of the dipole mode — 90
  - 5.1.2. Parametric excitation — 92
  - 5.1.3. Coupling Hamiltonian — 93
- 5.2. Measurement of atom loss in the surface potential — 95
  - 5.2.1. Determination of the cantilever position — 95
  - 5.2.2. Positioning reproducibility — 98
- 5.3. Model for surface induced atom loss — 99
  - 5.3.1. Improvements of the model — 100
  - 5.3.2. Heating rate analysis with the surface loss model — 101
- 5.4. Analysis of the surface potential — 103
  - 5.4.1. Additional potential U_{ad} — 103
  - 5.4.2. Iterative determination of the strength of U_{ad} — 104
  - 5.4.3. Adsorbates — 105
- 5.5. Detection of mechanical motion with BECs — 105
  - 5.5.1. Probing the cantilever fundamental mode spectrum — 106
  - 5.5.2. Distance dependence — 107
  - 5.5.3. Dependence on hold time — 108
- 5.6. Readout sensitivity — 109
- 5.7. Mode spectroscopy — 111
- 5.8. Simulation of cloud excitation — 114
  - 5.8.1. Simulation of classical trajectories — 114
  - 5.8.2. Simulation of 1D Gross-Pitaevskii dynamics — 118

6. Outlook — 123
- 6.1. Coupling BECs to a carbon nanotube — 123
- 6.2. Magnetic coupling of a BEC to a nanomechanical resonator — 127

6.3. Optical coupling of ultracold atoms to mechanical resonators	132
6.4. Conclusion	134

A. Fundamental constants and Rubidium Data   135

B. Fast trap ramping   136

C. Manoeuvre around the cantilever   138

D. Chip Fabrication   141

Bibliography   I

1. Introduction

Ultracold neutral atoms are an ideal system to study quantum physics in a very clean and directly accessible way. For the coherent manipulation and detection of the internal atomic state exists an elaborate toolbox [1], trapping techniques [2, 3] permit levitation and confinement of a gas, and Bose-Einstein condensation [4, 5, 6] facilitates the preparation of a cloud in the ground state of a trap, thereby enabling high control over the motional degrees of freedom [7, 8, 9]. Furthermore, coherent manipulation of internal states, control of interactions, and quantum non-demolition measurements enable the creation of non-classical states like single particle superpositions [10, 11, 9] or collective entangled states [12, 13, 14, 15]. High fidelity quantum manipulation of ultracold atoms is thus at hand.

A central advantage of neutral atomic quantum gases is their exceptional isolation from the environment. Their neutrality suppresses coupling to electric fields, and their magnetic field sensitivity can be suppressed by a proper choice of the used internal levels [10, 11] and by magnetic shielding. Finally, trapping of the gas in an UHV environment detains the atoms from any contact except for some infrequent collisions with the remaining background gas. With these conditions it becomes possible to achieve lifetimes of quantum states of several seconds [10, 11].

An important technique to facilitate the production and control of quantum gases was provided by the development of atom chips [16, 17, 18, 19]. The technology makes use of chip-based, microfabricated wires that generate versatile magnetic potentials for the trapping and transporting of ultracold atoms. The strong gradients which are achievable close to a wire provide substantially tighter trapping and thus faster production of Bose-Einstein condensates (BECs) compared to conventional techniques [20, 21], thereby also relaxing the vacuum requirements.

Beyond these technical advantages, atom chips have the potential to open a new perspective for research with ultracold gases. They enable the controlled trapping and versatile positioning of atoms close to surfaces. This gives the possibility of studying interactions between atoms and on-chip solid-state systems. Such interactions can be exploited on three different levels. First, atoms can be used as a sensitive local probe to detect electromagnetic fields and forces. This was e.g. beautifully demonstrated with measurements of the structure of the current density in evaporated gold [22, 23]. Other significant examples for this approach are precision measurements of surface forces [24, 25, 26, 27] or measurements of thermal magnetic near-field noise [28, 29, 24]. In a complementary approach, engineered solid state systems can be employed to manipulate and control atoms. This can either extend existing techniques or introduce novel concepts. A recent example is the coherent manipulation of BECs with microwave near-field radiation from a microfabricated waveguide on a chip [9]. Finally, extending the previous concepts to the quantum regime, the engineering of strong coupling between a solid sate system and an atomic system could enable coherent, bidirectional energy exchange on the single quantum level, very much as in the strong coupling limit of cavity quantum electrodynamics [30]. Such a coupling would provide a quantum interface between the two systems and enable e.g. coherent manipulation of the solid state system via the atoms. A variety of different systems has been considered in this context, including dipolar molecules [31] or neutral atoms [32] coupled to superconducting cavities, ions coupled to a mesoscopic electrode [33] or to a Cooper pair box [34], and atomic systems coupled to micro- and nanomechanical oscillators [35, 36, 37, 38, 39, 40, 41, 42, 43, 44, 45, 46, 47, 48].

Micro- and nanostructured mechanical oscillators [49, 50, 51] constitute particularly well suited coupling partners for such experiments. They are characterized by a spectrum of well resolved mechanical resonances, where mostly the fundamental mode is considered for experiments. It can serve as an isolated degree of freedom that is in many cases very well described by a simple harmonic oscillator weakly coupled to a thermal bath. Fundamental resonance frequencies range from kHz to GHz, such that possible couplings to atomic degrees of freedom may include atomic motional states, Zeeman transitions, and Hyperfine transitions. Their conceptual simplicity, moderate technical complexity (e.g. due to the possibility of room temperature operation), and the accessibility of the degree of freedom of interest distinguish them from other solid state systems. Micro- and nanomechanical resonators have attracted much attention lately, owing to the great achievements in the minimization of mechanical damping [52, 53, 54, 55], the improvement of readout sensitivity [56, 57, 58], the development of novel manipulation techniques for micromechanical motion [59, 60, 61], and the extreme sensitivity in force sensing applications [62, 63, 64]. One field that contributed much activity recently is cavity optomechanics [65, 66, 67, 68]. Its central accomplishment is the investigation of radiation pressure forces which allow one to manipulate the motional state of micromechanical oscillators. In particular, it has become possible to substantially cool the thermal excitation of a single mechanical mode, down to a few tens of remaining phonons [69, 70, 71, 72]. With these developments, micro- and nanomechanical resonators now represent an important model system with the prospect of demonstrating quantum effects on a macroscopic scale.

Interfacing ultracold atomic systems with mechanical resonators could both profit from and contribute to this development. The intriguing question is raised whether the sophisticated toolbox for coherent manipulation of the quantum state of atoms could be employed to read out, cool, and coherently manipulate mechanical oscillators. Several theoretical proposals have addressed this question and considered the coupling of micro- and nanomechanical oscillators to atoms [38, 39, 41, 42, 43, 44,

45, 46, 47, 48], ions [35, 36, 37], and molecules [40]. They show that sufficiently strong and coherent coupling would enable studies of atom-oscillator entanglement, quantum state transfer, and quantum control of mechanical force sensors.

In part of these scenarios the coupling relies on local field gradients, calling for very close approach of the atoms to the oscillator. In this respect, ground-state neutral atoms stand out because preparation [24] and coherent manipulation [11] at micrometer distance from a solid surface has already been demonstrated on atom chips. While the intrinsically weak coupling of neutral atoms to the environment enables long coherence times, it makes coupling to solid-state degrees of freedom non-trivial. So far, only first steps have been made to investigate coupling mechanisms experimentally, and the only experiment in this direction was performed in the group of John Kitching [73]. There, atoms in a heated vapor cell are coupled to a micromechanical cantilever with a magnetic tip. Piezo-excited oscillations of the cantilever create an oscillating magnetic field in the vapor cell and induce detectable spin precession. However, thermal motion of the atoms limits the interaction time and the control over the coupling.

This thesis

In this thesis I describe experiments realizing a controlled, resonant coupling between a micromechanical resonator and the collective motion of a Bose-Einstein condensate of ^{87}Rb atoms in a trap. We employ a novel coupling mechanism that does not require magnets, electrodes, or mirrors on the oscillator. The coupling is thus applicable to a large class of mechanical oscillators, including molecular-scale oscillators such as carbon nanotubes that are of particular interest due to their small mass.

The interaction relies on surface forces experienced by the atoms at about one micrometer distance from the mechanical structure. The forces deform the trapping potential and lead to the excitation of collective motion of the atoms in the trap when cantilever oscillations are resonant with a mechanical mode of the atoms, see Fig. 1.1. We observe resonant coupling to several well-resolved mechanical modes of the condensate, including in particular the center of mass mode and the breathing mode. The small spectral width of the atomic resonances provides an effective means to control the coupling via the trap frequency and e.g. permits to switch the coupling on and off. We use trap loss as the simplest way to detect BEC dynamics induced by the coupling. We can detect driven cantilever oscillations with a minimum resolvable amplitude of 13 nm within an interaction time of 3 ms. This value is limited by the strong anharmonicity of the trap, and by the short trap lifetime due to three-body collisional loss and technical heating. We show that the sensitivity could be improved by about two orders of magnitude by directly detecting the induced motion of the condensate.

To quantify the atom-cantilever interaction we develop a method to characterize

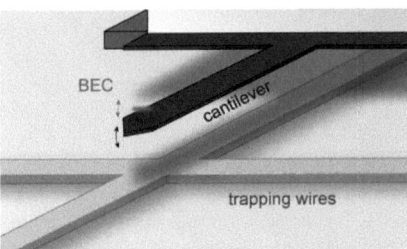

Figure 1.1.: Schematic setting of the experiment. We integrate a mechanical resonator on an atom chip. A Bose-Einstein condensate brought close to the cantilever is attracted by the surface potential and can be excited to collective motion via resonant cantilever oscillations.

the absolute strength of the surface forces with a combination of static loss and resonant coupling measurements. This allows us to precisely calibrate the atom surface distance and to infer the strength of the coupling constant that describes the interaction. The measurements are performed at a distance of $0.8 - 2.5$ μm from the surface, which requires exceptional precision in the positioning of the atoms. Using ultra-stable current sources, we achieve a positioning reproducibility below 6 nm rms, much less than the typical diameter of 500 nm of the condensates.

Application of this coupling mechanism to a single-wall carbon nanotube with improved readout of the atoms could permit to resolve the quantum fluctuations of the nanotube. Furthermore we discuss two alternative coupling schemes to achieve atom-cantilever coupling at the quantum level. Firstly, we describe the magnetic coupling of a nanomechanical resonator to the spin of a BEC. This scheme profits from the possibility of higher oscillator frequency, and from the long coherence lifetime and the high control over the spin degree of freedom. Secondly, we sketch a long distance coupling between a mechanical oscillator and laser cooled atoms via an optical lattice. Back action of the atoms on the mechanical oscillator becomes significant in these three scenarios, enabling manipulation and control of the oscillator. We discuss the conditions for which the strong coupling limit can be achieved.

Organization of the chapters

The **second chapter** gives an introduction to atom chips. I review the principle of magnetic microtraps and summarize the theory of Bose-Einstein condensation with a focus on collective excitations. A central section is then dedicated to surface forces and the resulting effects of a surface on trapped atoms nearby.

The **third chapter** introduces mechanical resonators and motivates their use in experiments on the coupling to ultracold atoms. I give an overview of detection and manipulation techniques of micro- and nanomechanical motion, and comment on the efforts to realize quantum states in these systems.

The **fourth chapter** describes the experimental setup and the techniques employed in the experiments. Furthermore, it summarizes measurements for the characterization of ultracold atoms close to a surface.

The **fifth chapter** presents the main results of this thesis. The coupling mechanism is explained, measurements for the determination of the surface potential are described, and measurements of dynamical atom-resonator coupling are presented and analyzed. Finally we present a numerical simulation that allows a quantitative interpretation of the data.

The **sixth chapter** gives an outlook on three different atom-resonator coupling scenarios that could be suited to study the coupling on the quantum level.

Work on fiber based Fabry-Perot resonators

Parallel to the studies described in this thesis I continued research on the development and the application of fiber based Fabry-Perot resonators. This work is covered in the publications listed below ([74, 75, 76, 77, 78]).

Contributions to publications

- *Stable fiber-based Fabry-Perot cavity* [74]
  T. Steinmetz, Y. Colombe, D. Hunger, T. W. Hänsch, A. Balocchi, R. Warburton, J. Reichel
  Appl. Phys. Lett. **89**, 111110 (2006).

- *Strong atom-field coupling for Bose-Einstein condensates in an optical cavity on a chip* [75]
  Y. Colombe, T. Steinmetz, G. Dubois, F. Linke, D. Hunger, J. Reichel
  Nature **450**, 06331 (2007).

- *Bose-Einstein condensate coupled to a nanomechanical resonator on an atom chip* [39]
  P. Treutlein, D. Hunger, S. Camerer, T. W. Hänsch, J. Reichel
  Physical Review Letters **99**, 140403 (2007).

- *Fluctuating nanomechanical system in a high finesse optical microcavity* [76]
  I. Favero, S. Stapfner, D. Hunger, P. Paulitschke, J. Reichel, H. Lorenz, E.

Weig, K. Karrai
Optics Express **17**, 12813 (2009).

- *Resonant Coupling of a Bose-Einstein Condensate to a Micromechanical Oscillator* [79]
  D. Hunger, S. Camerer, T. W. Hänsch, J. Reichel, D. König, J. P. Kotthaus, P. Treutlein
  Physical Review Letters **104**, 143002 (2010).

- *Fiber Fabry-Perot cavity with high finesse* [77]
  D. Hunger, T. Steinmetz, Y. Colombe, C. Deutsch, T. W. Hänsch, J. Reichel
  to be published in New Journal of Physics,
  preprint available on arXiv: 1005.0067 [physics.optics].

- CO_2 *laser fabrication of concave, low-roughness depressions on optical fibers end facets* [78]
  D. Hunger, C. Deutsch, R. Warburton, T.W. Hänsch, J. Reichel
  in preparation.

- *Optical Lattices with Micromechanical Mirrors* [47]
  K. Hammerer, K. Stannigel, C. Genes, M. Wallquist, P. Zoller, P. Treutlein, S. Camerer, D. Hunger, T. W. Hänsch
  preprint available on arXiv: 1002.4646 [quant-ph].

2. Atoms in magnetic chip traps

In this chapter I give an introduction to atoms in magnetic chip traps, the workhorse of the experiments described in this thesis. After covering the basics of trapping neutral atoms and the properties of magnetic microtraps, I will review the theory of ultracold and condensed quantum gases as it is important for the experiments presented in chapter 5. This includes mainly the description of atomic interactions, collective excitations, and effects of trap anharmonicity.

The central part of this chapter will then focus on the description of surface forces and their effect on trapped atoms near the surface. In chapter 5 we show that surface forces can be harnessed to realize a controlled, dynamical coupling between a mechanical resonator and ultracold atoms. Here we describe the origin of surface forces and their static effect on magnetic traps. We want to study the situation where atoms are brought as close as possible to a surface. Due to surface forces, the trapping potential will be deformed strongly and the remaining trap depth will be of the order of the atomic energy. In this regime, four effects, namely sudden loss, surface evaporation, tunneling, and quantum reflection affect the lifetime of the atomic cloud. Furthermore we discuss loss processes that limit experiments independent of the potential deformation. Especially collisional loss and technical heating become severe for the tight traps that are of interest for close approach to the surface.

We give quantitative models for the different loss processes that allow us to describe our experimental observations.

2.1. Magnetic microtraps

Magnetic traps are a standard tool to trap neutral atoms with magnetic moment. Due to the rather weak forces arising from magnetic interactions, the atoms have to be precooled before trapping is possible. The standard approach to realize a magnetic trap is to drive currents through macroscopic coils (10 cm scale) to generate magnetic field configurations for 3D enclosure. This bears the disadvantages of the need of large currents (typically 10^2 A), limited trap frequencies, and the restriction to simple, large scale geometries. These shortcomings can be overcome in an elegant way by using microfabricated wires to generate the magnetic potentials. Here one profits from the fact that a current I in a wire creates a magnetic field $B(r)$ and field gradient $B'(r)$ which scale as

$$B(r) = \frac{\mu_0}{2\pi} \frac{I}{r} \tag{2.1}$$

$$B'(r) = -\frac{\mu_0}{2\pi} \frac{I}{r^2}, \tag{2.2}$$

where r is the distance to the wire. Thus, by approaching the wire from cm distances to μm distances, one benefits from an increase of the magnetic field by a factor 10^4 and of the gradient by a factor 10^8. Thereby one can increase the confinement substantially, while the current requirements are at the same time relaxed to comfortable values of a few Ampére. Furthermore, the high achievable gradients are of major advantage for evaporative cooling, the standard technique to achieve Bose-Einstein condensation in trapped gases. This comes from the fact that the trap frequency, which is directly related to the field gradient, determines the speed of the cooling process. While for macroscopic magnetic traps with trap frequencies of a few hundred Hertz it can take more than a minute to reach BEC, microtraps with trap frequencies of several kHz can accomplish this within less than one second [80]. For such short experimental cycle times, background gas collisions become less important. In consequence, the vacuum requirements are relaxed and the apparatus can be simplified considerably (see chapter 4).

Besides these technical advantages, there are qualitatively new benefits from this approach: First, it gives the freedom to design quite arbitrarily shaped potentials of high complexity and scalability. This enables e.g. the creation of 1D potentials [81], double-well or even multi-well potentials [82], and a large range of trap aspect ratios. Furthermore, several structures such as wave guides, splitting junctions or "conveyor belts" can be combined on a single chip [83, 84, 19]. Finally, it provides techniques to manipulate and position ultracold atoms close to surfaces with high precision. Three dimensional controlled positioning above a surface is the central ingredient for studies of atom-surface interactions or controlled interfacing of atoms with solid state systems as studied in this thesis.

2.1.1. Magnetic trapping of neutral atoms

In the following we introduce the basics of magnetic trapping following Ref. [6]. Magnetic trapping is based on the Zeeman interaction of the magnetic moment $\boldsymbol{\mu}$ of a particle with an external magnetic field $\boldsymbol{B}(\boldsymbol{r})$. The classical interaction energy

$$E(\boldsymbol{r}) = -\boldsymbol{\mu} \cdot \boldsymbol{B}(\boldsymbol{r}) = -\mu B(\boldsymbol{r}) \cos\theta \tag{2.3}$$

depends on the angle θ between $\boldsymbol{\mu}$ and $\boldsymbol{B}$. When the magnetic field is inhomogeneous, the particle will feel a force proportional to the gradient of the field

$$\boldsymbol{f}(\boldsymbol{r}) = -\mu \nabla B(\boldsymbol{r}) \cos\theta, \tag{2.4}$$

2.1 Magnetic microtraps

where the direction of the force again depends on the relative orientation of the magnetic moment and the field. Minimization of the magnetic interaction energy will align $\boldsymbol{\mu}$ with $\boldsymbol{B}$ such that unlike poles face ($\theta = 0°$), and the force attracts the particle to the field maximum. The situation changes when the particle is rotating around the axis of $\boldsymbol{\mu}$. Then, θ is stabilized due to rapid Larmor precession of $\boldsymbol{\mu}$ around $\boldsymbol{B}$. This comes close to the quantum mechanical situation. For a quantum particle like an atom, the angle θ is quantized and the classical term $\cos\theta$ is replaced by the value m_F/F, where the magnetic quantum number m_F is the projection of the total angular momentum $\boldsymbol{F}$ on the direction of $\boldsymbol{B}$. The associated magnetic moment is then $\boldsymbol{\mu} = -\mu_B g_F \boldsymbol{F}$ and the quantum mechanical interaction energy reads

$$E_{F,m_F}(\boldsymbol{r}) = \mu_B g_F m_F B(\boldsymbol{r}), \tag{2.5}$$

where μ_B is the Bohr magneton and g_F is the Landé g-factor of the angular momentum state F. This position dependent energy describes a potential landscape for an atom in a certain m_F state. If $g_F m_F > 0$, the atom will be attracted to a magnetic field minimum and the state is called a "low-field seeker". As magnetic field maxima in free space do not exist according to Maxwell's equations, atoms have to be prepared in low-field seeking states for magnetic trapping.

Majorana spin flips

Trapping is only stable, if the atom remains in the initially prepared m_F state. This will be the case as long as the precessing spin can adiabatically follow the local direction of the magnetic field. Looking at the situation from the moving atom, this requires that the rate of change of the magnetic field direction θ is small compared to the precession frequency ω_L,

$$\frac{d\theta}{dt} \ll \omega_L(\boldsymbol{r}) = \frac{\mu_B |g_F| B(\boldsymbol{r})}{\hbar}. \tag{2.6}$$

If this condition is not fulfilled, transitions to "non-low-field seeking" states can occur, which lead to loss of the atom from the trap. These so called Majorana spin-flips can lead to a considerable loss rate in traps with vanishing (or small) minimum magnetic field strength $B_0 = \min(|\boldsymbol{B}|) = 0$ G. Calculations for anisotropic harmonic traps show that Majorana loss is suppressed exponentially for increasing minimum Larmor frequency [85], and a safe value for the magnetic field minimum of a trap is

$$B_0 \gtrsim 10 \frac{\hbar \omega}{\mu_B g_F}, \tag{2.7}$$

where we have assumed that $d\theta/dt$ equals the highest trap frequency ω in the trap.

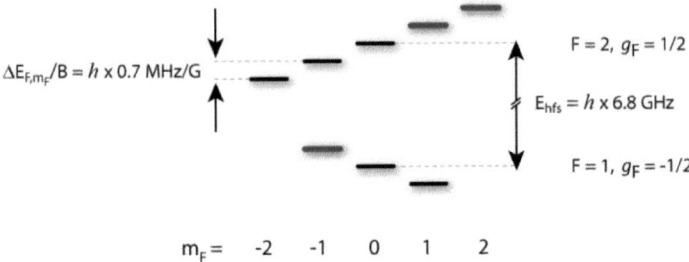

Figure 2.1.: Level scheme of the ground state of ^{87}Rb in a weak magnetic field. The Zeeman effect shifts the levels by $\Delta E_{F,m_F}$. Low field seeking states are $|1,-1\rangle, |2,1\rangle$ and $|2,2\rangle$.

Hyperfine structure

The above discussion describes an approximate potential for a ground state atom. It neglects that the total angular momentum $\boldsymbol{F} = \boldsymbol{I} + \boldsymbol{J}$ is composed of the nuclear angular momentum $\boldsymbol{I}$ and the electron angular momentum $\boldsymbol{J}$. The full Hamiltonian for a ground state atom in an external magnetic field is

$$H = A_{\text{hfs}} \boldsymbol{I} \cdot \boldsymbol{J} + \mu_B B (g_J J_z + g_I I_z), \tag{2.8}$$

where the first term describes the hyperfine interaction between the nuclear and electron angular momentum with the hyperfine structure energy A_{hfs}, and the other terms describe the couplings of $\boldsymbol{I}$ and $\boldsymbol{J}$ to the magnetic field, with the z–axis chosen parallel to $\boldsymbol{B}$. Note that $|g_I| \sim 10^{-3} \times g_J$ such that the coupling to the nuclear angular momentum leads only to a small energy contribution. The exact energy levels can be derived from diagonalizing the Hamiltonian 2.8, which for the case of $J = 1/2$ gives the Breit-Rabi formula [86]. For weak magnetic fields where E_{F,m_F} is small compared to the hyperfine energy, the deviation from equation 2.5 is small and can be neglected in many cases. For the $5^2 S_{1/2}$ ground state of ^{87}Rb the angular momenta are $J = 1/2$ and $I = 3/2$ so that F takes the two possible values $F = (1,2)$, and the hyperfine energy splitting is $E_{hfs} = A_{hfs}(I + 1/2) = h \times 6.8$ GHz. The Hyperfine energy levels of the ^{87}Rb ground state are shown in Figure 2.1.

2.1.2. Trap configurations

The two most basic types of magnetic traps can be discerned by the value of the field at the trap minimum. A quadrupole field has $B_0 = 0$ G and B rises linearly with the distance from the minimum. A Ioffe-Pritchard trap has a quadratic minimum with a non-zero value of B_0.

2.1 Magnetic microtraps

Quadrupole trap: The field of a quadrupole trap can be written as

$$\boldsymbol{B} = B'_x \boldsymbol{x} + B'_y \boldsymbol{y} + B'_z \boldsymbol{z} \tag{2.9}$$

where the field gradients have to satisfy $B'_x + B'_y + B'_z = 0$ to fulfill Maxwell's equations. For the case of $B'_z = 0$ the configuration is a 2D quadrupole in the xy−plane with $B'_y = -B'_x$. A 3D quadrupole can be created e.g. by a pair of coils in the Anti-Helmholtz configuration, where opposing currents are driven through two coaxial coils facing each other. This results in a field with gradients $B'_y = B'_z = -2B'_x$ with the common axis of the coils chosen along x.

Due to Majorana spin-flips, the zero crossing at the center of quadrupole traps effectively acts as a "hole" with radius $r \sim \sqrt{v\hbar/\mu B'}$, where v is the velocity of the atom. Such traps are useful for relatively hot clouds, where the atoms mostly populate orbits with large radii around the center. The minimum of a quadrupole trap can be shifted along all directions without changing the shape of the trap by superimposing a homogeneous field along the respective axis. This can be useful for transporting thermal ensembles over large distances (see chapter 4.3).

Ioffe-Pritchard trap: The field of a Ioffe-Pritchard trap has a finite magnetic field minimum and a quadratic confinement. It can be realized by superimposing a constant offset field B_0 which defines the trap axis, a 2D quadrupole field in the transverse plane to the trap axis, and a field with curvature along the trap axis [87]:

$$\boldsymbol{B} = B_0 \begin{pmatrix} 1 \\ 0 \\ 0 \end{pmatrix} + B' \begin{pmatrix} 0 \\ -y \\ z \end{pmatrix} + \frac{B''}{2} \begin{pmatrix} x^2 - (y^2 + z^2)/2 \\ -xy \\ -xz \end{pmatrix}. \tag{2.10}$$

Close to the minimum of the field configuration, the potential can be approximated by a radial symmetric, harmonic trap with trap frequencies

$$\omega_\perp = \sqrt{\frac{\mu_B g_F m_F}{m}} \frac{B'}{\sqrt{B_0}} \tag{2.11}$$

$$\omega_x = \sqrt{\frac{\mu_B g_F m_F}{m}} \sqrt{B''} \tag{2.12}$$

for atoms of mass m. Note that for the radial direction, the harmonic approximation is only valid for a small region around the center, and for larger distances the potential shows the linear dependence of the 2D quadrupole.

Wire traps

The two trap types can be realized with microfabricated, current carrying wires in many different ways. The first proposal of this kind was from Weinstein and Libbrecht [88], however requiring a rather complex wire geometry and several wires. The

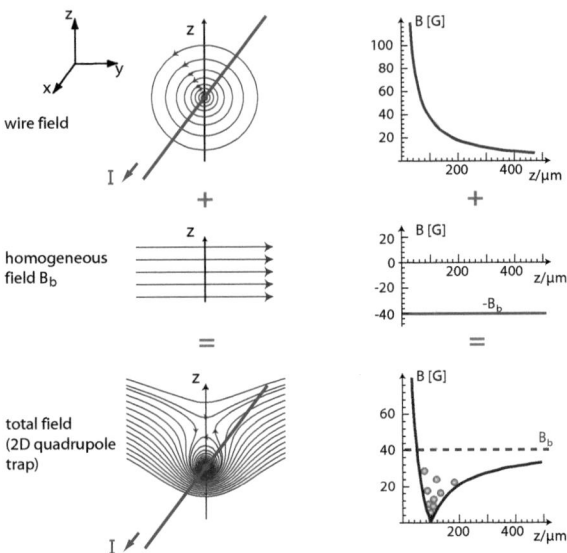

Figure 2.2.: Principle of the wire trap. Superposition of the radial field of a current in the wire with a homogeneous bias field in the plane transverse to the wire forms a 2D quadrupole guide. Left column: Magnetic field lines. Right column: Modulus of the magnetic field for a wire current $I = 2$ A and $B_{b,y} = 40$ G. Figure taken from [18].

following examples show the configurations which are commonly used in atomchip experiments.

Wire guide The basic building block of wire based magnetic traps is the wire guide. A DC current sent through a straight wire creates a circular magnetic field that decays according to equation 2.1 with $1/r$, where r is the distance to the wire. If a homogeneous field $B_{b,y}$ is superimposed perpendicular to the wire axis, a quadrupole field minimum is created at a distance

$$z_0 = \frac{\mu_0}{2\pi} \frac{I}{B_{b,y}} \qquad (2.13)$$

which forms a line parallel to the wire. This 2D confinement can already be used to guide atoms [89]. Figure 2.2 shows the field configuration.

2.1 Magnetic microtraps

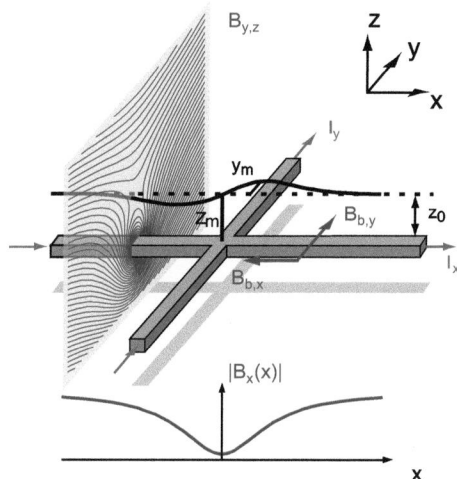

Figure 2.3.: The "Dimple" trap. A wire crossing with currents I_x, I_y and bias fields $B_{b,y}$ and $B_{b,x}$ allows to create a 3D Ioffe-Pritchard trap. Figure adapted from [16].

It is natural to fabricate the wires with microfabrication techniques on a substrate (see Chapter 4 and Appendix D). This provides the frame for a complex wire layout, and substrate materials with good heat conductivity allow to transport away the heat created by dissipation in the wires.

Ioffe-Pritchard Dimple trap To obtain 3D confinement, additional crossing wires can be added to create a configuration called Dimple trap. The schematic geometry is depicted in Fig. 2.3. It consists of a wire trap along e.g. the x-axis with a current I_x and a bias field $B_{b,y}$ as discussed above, which creates a 2D quadrupole in the yz-plane and defines the trap axis. A homogeneous field $B_{b,x}$ along the wire axis shifts the zero field minimum to a finite field value B_0 and creates a 2D Ioffe-Pritchard trap. To obtain axial confinement, the field from the current I_y in the crossing wire modulates the axial field. For small currents I_y in the crossing wire, the trap axis remains parallel to the main wire and the trap minimum position z_0 is determined only by I_x and $B_{b,y}$.

The parameters of the trap are then given by

$$B_0 = |B_{b,x} + \mu_0 I_y/2\pi z_0| \tag{2.14}$$
$$B'_z = \frac{\mu_0 I_x}{2\pi z_0^2} \tag{2.15}$$
$$B''_x = \mu_0 I_y/\pi z_0^3. \tag{2.16}$$

and the trap frequencies are approximated by

$$\omega_x = \sqrt{\mu/mB''_x}, \quad \text{and} \quad \omega_\perp = \sqrt{\frac{\mu}{m}\frac{B'^2_z}{B_0}}. \tag{2.17}$$

By designing arrays of wire crossings, an array of Ioffe-Pritchard traps can be created, where the geometry of each individual trap can be controlled independently. Such a configuration also enables to continuously shift the trap minimum along the trap axis to transport atoms.

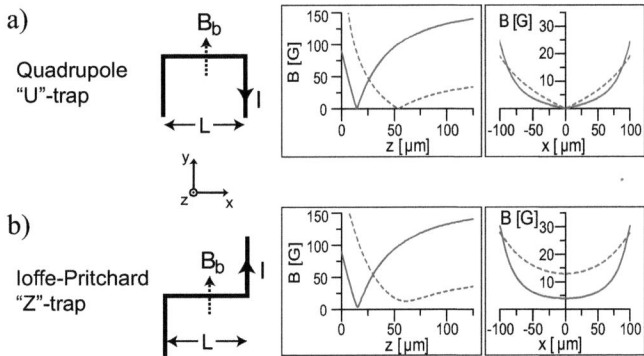

Figure 2.4.: Wire layout and magnetic field for a quadrupole "U"-trap (a) and an Ioffe-Pritchard "Z"-trap (b). Left column: wire layout in the plane z = 0 and orientation of the bias field. Center column: Magnetic field modulus on a line along z through the trap center. Right column: Magnetic field modulus on a line along x through the trap center. The fields were calculated for $L = 250$ μm and $I = 2$ A, taking a finite wire width of 50 μm into account. The bias field is $B_{b,y} = 54$ G (dashed lines) and $B_{b,y} = 162$ G (solid lines). Figure taken from [18].

Quadrupole U-trap and Ioffe-Pritchard Z-trap Alternatively to introduce additional crossing wires, a single wire can also be bent to achieve 3D confinement. The central part of the wire is used to create a 2D quadrupole and to define the trap axis in the same manner as a wire trap. The necessary field components along the trap axis are provided by the bent sections of the wire, orthogonal to the central part. When bending the wire in a "U" shape, the magnetic fields of the two bent sections are opposing each other and cancel in the center of the trap. This provides a 3D quadrupole trap. When bent in a "Z" shape, the magnetic fields add up to a finite field at the trap center and thus provide the parabolic confinement of a Ioffe-Pritchard trap. Figure 2.4 shows the wire configuration and the magnetic fields along the z-axis pointing out of the plane of the wire, and along the trap axis parallel to the x-axis.

2.2. Properties of Bose-Einstein condensates

One great achievement of experiments with ultracold atoms is the full control over the quantum state of the gas. Preparation of all atoms in one internal state by optical pumping, and the preparation of the ensemble in the motional ground state by Bose-Einstein condensation results in a system in a well initialized single quantum state. It has been identified that collective mechanical modes of a BEC can serve as a perfectly isolated mechanical oscillator [90, 91], which permits to study quantum mechanics and measurement back action in a very clean system. The first experiments along this line [90, 91, 92] explore the mechanics of BECs in optical cavities. In our experiments, we use collective excitations to directly probe mechanical motion of a mechanical resonator.

In this section we describe the theory of dilute, weakly interacting Bose-Einstein condensates and review two different limits to describe condensate excitations.

Bose-Einstein condensation (BEC) was predicted already in 1924 by Albert Einstein [93], based on ideas from Satyendra Bose. The first realization of this state of matter was demonstrated with the superfluidity of liquid ^{4}He. However, in this system the atoms are strongly interacting and only $\sim 10\%$ of the atoms populate the ground state. In dilute atomic gases, BEC was first experimentally realized in 1995 [4, 5, 94]. Here, interactions are weak and very pure condensates can be created. Condensation occurs at sub-microkelvin temperatures. At such low temperatures, inelastic atomic collisions lead to stable molecules, and the gaseous phase is only a metastable configuration. An important condition for the possibility of condensation is thus that the rate of elastic collisions, which are responsible for the thermalization of the gas, is much larger than the rate of inelastic collisions, which lead to trap loss and molecule formation.

The phenomenon of BEC is the direct consequence of the statistics of an ensemble of undistinguishable bosonic particles [95]. It occurs as a non-trivial phase transition when a gas of atoms is cooled down to ultralow temperatures. The remarkable point is that already at finite temperature, below a certain critical value T_c, *bosonic stimulation* leads to a scattering of atoms from thermally excited states to the ground state [96]. This leads to a macroscopic occupation of the ground state and results in a markedly non-thermal energy distribution. Described in position space, condensation occurs when the size of an atomic wave packet, given by the thermal deBroglie wavelength $\lambda_{dB} = \sqrt{2\pi\hbar^2/mk_BT}$, becomes larger than the interatomic spacing $n_0^{1/3}$, such that the wavepackets overlap and the atoms loose their identity. Here, m is the mass of the atom, n_0 the maximum density, and T the temperature of the cloud. More accurately, for a three dimensional uniform gas in the thermodynamic limit, the condition for condensation is given by

$$n_0 \lambda_{dB}^3 \geq 2.61. \tag{2.18}$$

The product in Eq. 2.18 equals the phase space density and is directly proportional to the atom number in the condensate. A condensate is thus an object composed of a macroscopic number of particles that share a single quantum state which can be described by a single wave function.

The temperature at which the phase transition occurs for a trapped cloud can be calculated analytically [95], and for a gas in the thermodynamic limit one obtains

$$T_c = 0.94 \frac{\hbar \omega_{\text{ho}}}{k_B} N^{1/3}, \tag{2.19}$$

where N is the total atom number and $\omega_{\text{ho}} = (\omega_x \omega_y \omega_z)^{1/3}$ denotes the geometric average of the trap frequencies along the the main axes of the trap. For $T < T_c$ the number of particles in the condensate is

$$N_0(T) = N[1 - (T/T_c)^3], \tag{2.20}$$

such that for $T \ll T_c$ effectively all atoms are in the ground state. For small atom number, as it is the case in the experiments described here, the finite size of the cloud leads to a smaller condensate fraction [95]

$$N_0(T) = N \left[1 - \left(\frac{T}{T_c}\right)^3 - 2.18 \frac{\bar{\omega}}{\omega_{\text{ho}}} \left(\frac{T}{T_c}\right)^2 N^{-1/3} \right], \tag{2.21}$$

with the ratio of the arithmetic ($\bar{\omega}$) and geometric (ω_{ho}) averages of the trap frequency.

2.2 Properties of Bose-Einstein condensates

2.2.1. Gross-Pitaevskii description

Binary elastic collisions between atoms play an important role for both static and dynamic properties of a condensate. In general, collisions depend strongly on the interatomic scattering potential U_{int}, which is often approximated by a Lennard-Jones type potential $U_{\text{int}} = C_{12}/r^{12} - C_6/r^6$. However, at low temperature, when the deBroglie wavelength is much larger than the effective range of the interatomic potential, collisions can be described in a simplified manner by a hard sphere contact potential

$$U_{\text{int}}(\boldsymbol{r}_i - \boldsymbol{r}_j) = \frac{4\pi\hbar^2 a_s}{m}\delta(\boldsymbol{r}_i - \boldsymbol{r}_j) \tag{2.22}$$

with the s-wave scattering length a_s. The coupling constant $g = 4\pi\hbar^2 a_s/m$ characterizes the strength of the collisional interaction. In the case of ^{87}Rb, the interaction is repulsive and $a_s = 5.0$ nm [97, 98] for the state $|2,2\rangle$.

In second quantization, one describes the condensate by a bosonic field with particle creation and annihilation operators $\hat{\psi}^\dagger(\boldsymbol{r}), \hat{\psi}(\boldsymbol{r})$, and the Hamiltonian for the system in an external potential U_{ext} is given by

$$\hat{H} = \int d\boldsymbol{r}\hat{\psi}^\dagger(\boldsymbol{r})\left(-\frac{\hbar^2}{2m}\nabla^2 + U_{\text{ext}}\right)\hat{\psi}(\boldsymbol{r}) + g\int d\boldsymbol{r}\hat{\psi}^\dagger(\boldsymbol{r})\hat{\psi}^\dagger(\boldsymbol{r})\hat{\psi}(\boldsymbol{r})\hat{\psi}(\boldsymbol{r}). \tag{2.23}$$

The first part of Eq. 2.23 represents the kinetic and potential energy, while the last term accounts for the collisional interactions, where we have made use of Eq. 2.22. Despite the high phase space density, BECs are dilute gases in the sense that the characteristic size a_s of the atom is much smaller than the mean interatomic distance $\langle n \rangle^{1/3}$, with $\langle n \rangle$ the mean density. In this limit, $\langle n \rangle a_s^3 \ll 1$, the collisional interaction can be approximated by a mean field potential felt by each atom. In the mean field description, the bosonic field operator $\hat{\psi}(\boldsymbol{r})$ is decomposed into an expectation value $\Phi(\boldsymbol{r}) = \langle \hat{\psi}(\boldsymbol{r}) \rangle$ that describes the condensate and which is called the order parameter, and a field operator $\hat{\Psi}(\boldsymbol{r})$ that describes excitations,

$$\hat{\psi}(\boldsymbol{r}) = \Phi(\boldsymbol{r}) + \hat{\Psi}(\boldsymbol{r}). \tag{2.24}$$

The order parameter represents a macroscopic wave function, which is just the renormalized single particle wave function $\phi(\boldsymbol{r})$ into which condensation occurs,

$$\Phi(\boldsymbol{r}) = \sqrt{N_0}\phi(\boldsymbol{r}). \tag{2.25}$$

Here, N_0 is the number of condensate atoms with a density distribution $n_c(\boldsymbol{r}) = |\Phi(\boldsymbol{r})|^2$.

In the simplest approximation, the excitations $\hat{\Psi}(\boldsymbol{r})$ are neglected, and one obtains the well known Gross-Pitaevskii equation (GPE)

$$\left(-\frac{\hbar^2}{2m}\nabla^2 + U_{\text{ext}}(\boldsymbol{r}) + gN_0|\phi(\boldsymbol{r})|^2\right)\phi(\boldsymbol{r}) = \mu_c\phi(\boldsymbol{r}) \tag{2.26}$$

which determines the wave function $\phi(\boldsymbol{r})$ and the condensate chemical potential μ_c. It is valid for low temperatures $T \ll T_c$, large atom number $N_0 \gg 1$, and weak interaction $\langle n \rangle a_s^3 \ll 1$.

The dynamics of the condensate wavefunction is given by the time-dependent Gross-Pitaevskii equation

$$i\hbar \frac{\partial}{\partial t}\phi(\boldsymbol{r},t) = \left(-\frac{\hbar^2}{2m}\nabla^2 + U_{\text{ext}}(\boldsymbol{r}) + gN_0|\phi(\boldsymbol{r},t)|^2\right)\phi(\boldsymbol{r},t). \tag{2.27}$$

In the mean field description, a new length scale enters, the healing length ξ of the condensate

$$\xi = \hbar/\sqrt{2mgn}. \tag{2.28}$$

It is the minimal distance, over which the condensate wave function can vary significantly.

Thomas-Fermi approximation

If the kinetic energy is much smaller than the interaction and potential energy $\hbar\omega_{\text{ho}} \ll gn_c$, a simple solution of the GPE can be found by neglecting the kinetic energy term in Eq. 2.26. It is known as the Thomas-Fermi approximation, which results in simple analytical formulas for the main properties of a BEC:

$$n_c(\boldsymbol{r}) = |\phi(\boldsymbol{r})|^2 = \max\{0, [\mu_c - U_{\text{ext}}(\boldsymbol{r})]/g\} \tag{2.29}$$

$$\mu_c = \frac{\hbar\omega_{\text{ho}}}{2}\left(\frac{15a_s N_0}{a_{\text{ho}}}\right)^{2/5} \tag{2.30}$$

$$R_{\text{TF},i} = \sqrt{2\mu_c/m\omega_i^2} \qquad i = x, y, z. \tag{2.31}$$

The density profile $n_c(\boldsymbol{r})$ is directly given by the shape of the trapping potential rather than by the single particle ground state wave function (a gaussian for a harmonic trap), and repulsive interaction leads to larger cloud radii $R_{\text{TF},i}$, called the Thomas-Fermi radii. Due to the interaction, also the chemical potential is larger than the ground state energy by the Thomas-Fermi factor $\chi_{TF} = (15N_0 a_s/a_{\text{ho}})^{2/5}$, where $a_{\text{ho}} = \sqrt{\hbar/m\omega_{\text{ho}}}$ is the mean oscillator length. In our experiments, condensates with small atom number ($N_0 \sim 10^3$) are prepared, and the TF-limit $Na_s/a_{\text{ho}} \gg 1$ is not always satisfied. An interpolation between the TF-regime and the non-interacting regime $N_0 a_s/a_{\text{ho}} < 1$ can be used to calculate the precise cloud properties in this case [99, 100].

2.2.2. Condensate excitations

In the GPE, only the condensate wave function and the interaction within the condensate is described. To cover excitations and interaction induced correlations, the

2.2 Properties of Bose-Einstein condensates

field operator $\hat{\Psi}(r)$ for the excitations has to be included. There is a hierarchy of approximations regarding how the excitations are treated. In the following, we review two relevant limits: First the Bogoliubov theory, which gives predictions for collective mode frequencies that we observe in our experiments. However, this description neglects the thermal component of the gas, which is important e.g. for surface loss and for the damping of excitations. To include these effects, we discuss the Popov approximation in the second section.

Bogoliubov theory

The simplest approximation is the Bogoliubov theory: It considers excitations as small fluctuations with vanishing expectation value $\langle \delta \hat{\Psi} \rangle = 0$. This corresponds to the zero temperature limit, where the fluctuations are not thermally populated.

The resulting Hamilton operator can be diagonalized when the field operator $\hat{\Psi}$ is written in a form given by the Bogoliubov transformation

$$\hat{\Psi}(r) = \sum_k u_k(r)\hat{b}_k - v_k(r)\hat{b}^\dagger_{-k} \qquad (2.32)$$
$$\hat{\Psi}^\dagger(r) = \sum_k u_k(r)\hat{b}^\dagger_{-k} - v_k(r)\hat{b}_k$$

with the "quasiparticle" and "hole" mode functions $u_k(r), v_k(r)$ of wave vector k, and the creation and annihilation operators $\hat{b}^\dagger, \hat{b}$ of bosonic quasiparticle excitations.

The solution of the Hamilton operator results now in three equations. The first is the GPE (Eq. 2.26) for the condensate, which remains unaffected from the excitations. The remaining two equations are for the mode functions of the quasiparticles, the coupled Bogoliubov equations [95]

$$\hbar \omega_k u_k(r) = \left(\mathcal{H}_0 + 2gN_0\phi^2(r)\right) u_k(r) + gN_0\phi^2(r)v_k(r), \qquad (2.33)$$
$$-\hbar \omega_k v_k(r) = \left(\mathcal{H}_0 + 2gN_0\phi^2(r)\right) v_k(r) + gN_0\phi^2(r)u_k(r),$$

with $\mathcal{H}_0 = -(\hbar^2/2m)\nabla^2 + U_{\text{ext}}(r)$. The solution of these equations gives the excitation spectrum of the condensate. However, analytic solutions for a trapped condensate only exist for the special case of spherical symmetry of a harmonic trap [101] or in the homogeneous case [102]. For a homogeneous gas, the amplitudes of the quasiparticle modes satisfy [102]

$$u_k^2 = v_k^2 + 1 = \frac{1}{2}\left(\frac{\epsilon_k^0 + gn}{\epsilon_k} + 1\right) \qquad (2.34)$$

with ϵ_k the Bogoliubov quasiparticle energy

$$\epsilon_k = \sqrt{\epsilon_k^0(\epsilon_k^0 + 2gn)} = gn\sqrt{(k\xi)^2[(k\xi)^2 + 2]}, \qquad (2.35)$$

and the free particle energy $\epsilon_k^0 = \hbar^2 k^2 / 2m$.

The equations describe excitations that are composed of atoms moving with $\pm \hbar k$, where u_k^2 gives the number of atoms moving in the direction of the excitation and v_k^2 atoms moving in the opposite direction. Although typically large numbers of atoms contribute to an excitation, the net momentum carried by a single excitation is equal to $\hbar k$.

Furthermore, Eq. 2.35 describes a quasiparticle spectrum with two regimes: For low energy excitations with $k\xi < 1$, where the excitation wavelength $\sim k^{-1}$ is larger than the healing length ξ, the condensate can deform on the scale of the wave vector and the excitations propagate like a sound wave in a medium. This results in a phonon like, linear dispersion

$$\epsilon_k \approx ck \tag{2.36}$$

where $c = \sqrt{gn/m}$ is the speed of sound of the condensate. In the phonon regime, the amplitudes $u_k^2, v_k^2 \approx 1/2k\xi > 1$ describe collective excitations with many atoms contributing to the mode.

For high energy excitations with $k\xi > 1$, the condensate wavefunction can not adapt to the modulation, and the spectrum becomes single-particle like with a quadratic dispersion relation

$$\epsilon_k \approx \frac{\hbar^2 k^2}{2m} + gn. \tag{2.37}$$

In the single-particle regime, $u_k^2 \approx 1$ and $v_k^2 \approx 0$, and there is an extra energy gn required for the excited atoms to move with $k\xi > 1$ in the surrounding gas.

Collective mode frequencies We now discuss the excitations of a trapped, repulsively interacting gas. When the interaction energy gn is large and the Thomas-Fermi approximation is valid, equations 2.33 coincide with the hydrodynamic equations for superfluids. In the spherical case, the eigenfrequencies of low frequency excitations have the analytic form [101]

$$\omega(n, \ell) = \omega_{\text{ho}} \sqrt{2n^2 + 2n\ell + 3n + \ell}, \tag{2.38}$$

where n and ℓ are the principal and the angular momentum quantum number, respectively. This is in contrast to the non-interacting case, where $\omega(n, \ell) = \omega_{\text{ho}}(2n + \ell)$.

The result can also be extended to the case, where the kinetic energy E_{kin} is not negligible compared to the potential energy E_{pot} (as assumed in the TF approximation). For the often studied quadrupole mode ($n = 0, \ell = 2$), which is also important for our experiments, the mode frequency is

$$\omega_{0,2} = \sqrt{2}\omega_{\text{ho}} \sqrt{1 + E_{\text{kin}}/E_{\text{pot}}}. \tag{2.39}$$

2.2 Properties of Bose-Einstein condensates

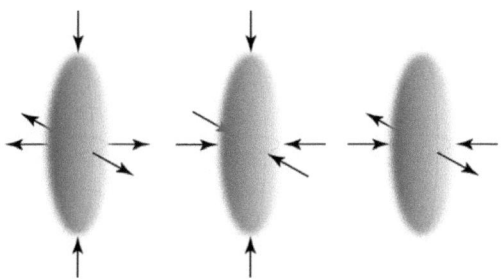

Figure 2.5.: Three of the lowest collective excitations of a BEC in a prolate trap $\omega_\perp \gg \omega_x$. Left: Low frequency ($m = 0$, $\ell = 2$) quadrupole mode with frequency $\omega = \sqrt{5/2}\,\omega_x$. Middle: High frequency ($m = 0, \ell = 0$) radial compression mode with frequency $\omega = 2\omega_\perp$. Right: Quadrupole mode ($m = 2$, $l = 2$) with frequency $\omega = \sqrt{2}\,\omega_\perp$.

For the case of a trap with cylindrical symmetry as used in our experiments, the generalization to prolate clouds with $\omega_\perp \gg \omega_x$ has to account for the dependence on the axial component of the angular momentum, the quantum number m. One obtains

$$\omega^2_{\ell,m=\pm\ell} = \ell\omega^2_\perp \qquad (2.40)$$
$$\omega^2_{\ell,m=\pm(\ell-1)} = (\ell - 1)\omega^2_\perp + \omega^2_x \qquad (2.41)$$

For the quadrupole mode ($\ell = 2$) with $m = 0$, a coupling to the monopole mode ($n = 1, \ell = 0$) leads to two decoupled modes with frequencies

$$\omega_{\ell=2,m=0} = \begin{cases} \sqrt{5/2}\,\omega_x \\ 2\,\omega_\perp \end{cases} \qquad (2.42)$$

in the limit of $\omega_x \ll \omega_\perp$. Figure 2.5 depicts the major low lying collective modes.

For small clouds beyond the TF regime, where the kinetic energy has to be included, the quadrupole mode frequency is calculated similarly to Eq. 2.39 and reads [103]

$$\omega_{2,2} = \sqrt{2}\omega_\perp \sqrt{1 + E_{\text{kin},\perp}/E_{\text{pot},\perp}}. \qquad (2.43)$$

The kinetic and potential energy along the transverse dimension $E_{\text{kin},\perp}$, $E_{\text{pot},\perp}$ can be evalutated following Ref. [100].

In experiments, collective oscillations were the first object of study after demonstration of BEC. Early experiments on low lying collective modes were demonstrated in [104, 105, 106]. More recently, a high order collective mode was excited by a cavity standing wave, and the dynamical coupling between the cavity light field and the collective mechanical motion was studied [91, 107]. In our experiments we observe the excitation of low lying collective modes by the coupling to an oscillating micro-cantilever (see chapter 5).

Popov approximation

In contrast to the Bogoliubov approximation, the Popov approximation includes a thermally occupied quasiparticle spectrum, and neglects only the so-called anomalous density term $\tilde{m} = \langle \hat{\Psi}(r)\hat{\Psi}(r) \rangle$ and terms with higher than quadratic order in $\hat{\Psi}$. This means that besides the condensate density $n_c(r) = |\Phi(r)|^2$ one now also solves for a thermal density distribution $n_T = \langle \hat{\Psi}^\dagger \hat{\Psi} \rangle$. The condensate wave function is determined by the generalized GPE [108, 109]

$$[\mathcal{H}_0 + gn_c(r) + 2gn_T(r)]\,\Phi(r) = \mu_c \Phi(r). \tag{2.44}$$

The thermal cloud is obtained by a Bogoliubov transformation (Eqs. 2.32) of $\hat{\Psi}$ which leads to coupled Popov equations similar to Eqs. 2.33, now also containing a mean field contribution due to n_T. A solution for the quasiparticle energies ϵ_k and amplitudes u_k, v_k determines the non-condensate density

$$n_T(r) = \sum_k \frac{|u_k(r)|^2 + |v_k(r)|^2}{\exp(\epsilon_k/k_B T) - 1} + |v_k(r)|^2. \tag{2.45}$$

The first term is the thermal occupation of the quasiparticle states and describes the "normal" thermal cloud. The last term gives a finite thermal density distribution also for $T = 0$ and is called quantum depletion. This accounts for the fact that collisions within the condensate can scatter atoms to excited states. For dilute gases with weak interactions, quantum depletion is of the order $(N - N_c)/N = (8/3)\sqrt{n_0 a_s^3/\pi}$, eg. for the case of ^{87}Rb and maximum density $n_0 = 10^{15}$ /cm a depletion of 1%. This is in contrast to suprafluid ^{4}He, where quantum depletion limits the condensate fraction to $\sim 10\%$.

Note that due to atom bunching in the thermal cloud, the mean field interaction for thermal atoms is twice as large as for condensate atoms. Thus, while for condensate atoms the effective potential becomes flat in the TF limit, there will be a bump for thermal atoms. This pushes the thermal component away from the trap center, forming a thermal shell around the condensate. On the other hand, this shell contributes to an effective potential for the condensate and compresses it [110]. The density distributions for the thermal and condensed cloud have to be found self-consistently to fulfill equations 2.44 and the generalizations of Eqs. 2.33.

In the semi-classical Thomas-Fermi approximation, the densities are given by two

2.2 Properties of Bose-Einstein condensates

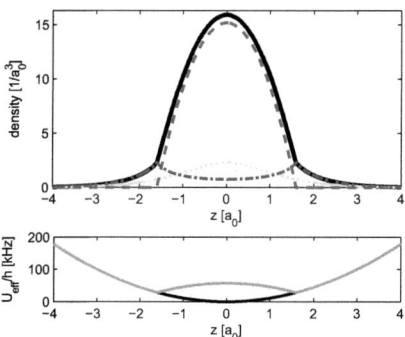

Figure 2.6.: Top: Bimodal density distribution (cut along the radial direction). Dashed line: n_c, dotted: undisturbed thermal cloud, dash-dotted: n_T, solid: $n = n_c + n_T$. Bottom: Effective potential U_{eff} for the thermal cloud in the presence of a TF-condensate. The thermal cloud sees the meanfield of the condensate density as a bump in the effective potential and is thereby repelled from the trap center. Calculation for 2000 atoms in a 10 kHz trap at $T = 0.7\, T_c$ with condensate fraction of 60% in dimensionless units $a_0 = \sqrt{\hbar/m\omega_\perp}$.

coupled equations [108]

$$n_T(\mathbf{r}) = \lambda_{dB}^{-3} g_{3/2} \left[\exp\left(-\frac{2gn_T(\mathbf{r}) + 2gn_c(\mathbf{r}) + U_{\text{ext}}(\mathbf{r}) - \mu_c}{k_B T} \right) \right], \quad (2.46)$$

$$n_c(\mathbf{r}) = \max\left\{ \frac{\mu_c - 2gn_T(\mathbf{r}) - U_{\text{ext}}}{g}, 0 \right\}, \quad (2.47)$$

where $g_{3/2}(z) = \sum_l z^l/l^{3/2}$ is the polylogarithm function. The mean field potential arising from the thermal cloud $2gn_T(\mathbf{r})$ is usually much smaller than the condensate mean field and can be neglected in many cases. In this limit, the two equations become decoupled and the density profiles can be obtained easily. Figure 2.6 shows the density distribution of a partially condensed cloud at $T = 0.7\, T_c$ with $N = 2000$ in a trap with $\omega_\perp = 10$ kHz, $\omega_x = 1$ kHz where the condensate fraction is determined by Eq. 2.21 and the themal cloud mean field is neglected.

The Popov approximation describes all important effects properly for temperatures far enough below T_c. In particular the repulsion of the thermal cloud due to the condensate mean field, damping of condensate excitations due to friction by interaction with the thermal cloud, and excitation of the condensate by thermal cloud motion or vice versa are predicted.

However, the Popov approximation fails to explain experiments with large thermal fractions ($> 50\%$). This situation is described correctly in the Hartree-Fock-

Bogoliubov theory, where only terms of $\hat{\Psi}(r)$ and $\hat{\Psi}^\dagger(r)$ in fourth order are neglected. The main extension is the inclusion of the coupling between excitations and the so-called anomalous density, which describes correlations between thermal atoms. This theory also explains excitations for temperatures close to T_c with good accuracy, and e.g. describes the unexpected large damping of collective excitations at temperatures close to T_c [111, 112].

2.2.3. Condensate motion in anharmonic traps

A collective mode of a condensate represents a very well isolated mechanical oscillator which is initially prepared in its ground state [90, 91, 92]. A special role is taken by the lowest excitation of the cloud, the center of mass (c.o.m.) or dipole mode. The c.o.m. motion of an ensemble of atoms in a harmonic trap leaves the internal cloud dynamics unaffected, regardless of the strength of interactions within the system. This is a result of the Kohn theorem [113]. As a consequence, the dipole mode of a cloud can act as a mechanical oscillator with very high quality factor, which is only limited by the lifetime of the atoms in the trap ($Q \sim 10^4$ was demonstrated in [26]).

However, it was mentioned in the previous section, that condensate excitations are subject to considerable damping when the thermal component does not oscillate in phase. Due to the different density distributions and the discrepancy in speed of sound, this will be the case for all modes but the dipole mode in a harmonic potential. For this mode, the condensate and the thermal cloud oscillate together, such that the center of mass coordinate of the total density distribution

$$\boldsymbol{R}(t) = \frac{\int \boldsymbol{r} n(\boldsymbol{r}, t) d^3 r}{\int n(\boldsymbol{r}, t) d^3 r} \tag{2.48}$$

performs oscillations of the form

$$R_i(t) = \sum_i a_i \sin(\omega_i t + \theta_i), \qquad i = x, y, z \tag{2.49}$$

and the shape of the cloud remains unchanged, independent of interactions and mean field potentials associated with $n(\boldsymbol{r})$.

Trap anharmonicities change the situation and can lead to excitation of higher modes and thereby cause damping. We give an analytical description for the GPE dynamics in potentials with small anharmonicity following [114, 115, 116] and [117] where the thermal cloud is neglected for simplicity. (An alternative approach to describe excitation in an anharmonic trap was given by [118].)

The solution of the time-dependent GPE 2.27 can be written in the form

$$\phi(\boldsymbol{r}, t) = \phi_0(\boldsymbol{r} - \boldsymbol{R}, t) e^{iS(\boldsymbol{r})} e^{i\varphi(t)} \tag{2.50}$$

2.2 Properties of Bose-Einstein condensates

with a phase S whose gradient is given by the velocity of the c.o.m. coordinate

$$\nabla S(\mathbf{r}, t) = \frac{m}{\hbar} \dot{\mathbf{R}} \qquad (2.51)$$

and an additional global phase $\varphi(t)$. The dynamics of the c.o.m. coordinate can be described by the classical equation of motion

$$m\ddot{\mathbf{R}} = -\nabla U_{\text{ext}}(\mathbf{R}), \qquad (2.52)$$

and the in general anharmonic potential U_{ext} can be expanded in a taylor series around the c.o.m. coordinate of the cloud.

For small anharmonicity, the potential can be approximated as locally harmonic, and it is sufficient to expand the series to second order. In the frame of the condensate moving in this potential, the anharmonicity appears as a time-dependent oscillation frequency

$$\omega_i^2(t) = \frac{1}{m} \frac{\partial^2}{\partial r_i^2} U_{\text{ext}}(\mathbf{R}(t)), \qquad i = x, y, z. \qquad (2.53)$$

In the TF limit, an exact analytical solution can be found for this problem. It describes the time evolution of the parabolic density distribution by rescaling factors $\lambda_i(t)$ of the TF-radii [114]

$$n_c(\mathbf{r}, t) = \frac{\mu_c}{g} \frac{1}{\lambda_x(t)\lambda_y(t)\lambda_z(t)} \left[1 - \sum_{i=x,y,z} \left(\frac{r_i}{R_{\text{TF},0i}\lambda_i(t)} \right)^2 \right]. \qquad (2.54)$$

The rescaling factors determine the complete time evolution of the TF radii $R_{\text{TF},i}(t) = \lambda_i(t) R_{\text{TF},0i}$ and satisfy the differential equations

$$\ddot{\lambda}_i = \frac{\omega_{0i}^2}{\lambda_i \lambda_x \lambda_y \lambda_z} - \omega_i^2(t) \lambda_i, \qquad i = x, y, z \qquad (2.55)$$

In this way, the anharmonicity couples the c.o.m. motion and the time evolution of the TF radii. Experimentally such a coupling to the cloud shape has been demonstrated with large amplitude oscillations ($b \approx 1$ mm) of a BEC in an anharmonic wave guide [115]. The two $l = 2, m = 0$ modes at $\sqrt{5/2}\omega_x$ and $2\omega_x$, and nonlinear mode mixing was observed.

To describe the situation of our experiments, where anharmonicity can be very strong, this analytical description is of limited use and we employ a numerical simulation of the GPE to investigate the cloud dynamics (see chapter 5.8.2).

2.3. Surface forces

In the following we direct our view on the situation of an ultracold cloud or a BEC in ultimate proximity to a surface. In such a setting, the presence of surface forces has an important influence on the behaviour of the atoms. We discuss different origins of surface potentials and show that they are typically attractive and decay over a characteristic length of a few microns.

2.3.1. Van der Waals-London and Casimir-Polder Potential

A neutral ground state atom close to a surface feels an attractive force. It arises due to the interaction of the fluctuating electric dipole of the atom with the electromagnetic field, which is modified by the presence of the surface. Depending on the atom-surface distance, three different regimes can be distinguished [119].

Van der Waals-London regime

For distances z where the propagation time of electromagnetic waves at the frequency of the strongest atomic transition can be neglected, the force can be described as the interaction of a dipole with its image dipole. The fluctuating electric dipole moment causes a fluctuating image dipole, which leads to an attractive interaction, called the London dispersion force or one type of van der Waals force. Near a dielectric surface with refractive index $n_r = \sqrt{\epsilon_r}$, the potential is described by

$$U_{\text{vdW}}(z) = -\frac{\hbar}{16\pi^2 \epsilon_0} \frac{1}{z^3} \int_0^\infty \alpha(i\xi) \frac{\epsilon_r(i\xi) - 1}{\epsilon_r(i\xi) + 1} d\xi \equiv -\frac{C_3}{z^3} \qquad (2.56)$$

with $a(i\xi)$ the polarizability of the atom evaluated for imaginary frequencies. For a perfectly reflecting surface, $\epsilon_r \to \infty$ and the fraction in the integral becomes unity.

Casimir-Polder regime

For distances larger than $\lambda/2\pi$, retardation of the electromagnetic field becomes important, and a QED description of the situation is necessary. Casimir and Polder [120] derived the potential for a perfectly conducting surface

$$U_{\text{CP}}(z) = -\frac{3\hbar c}{32\pi^2 \epsilon_0} \frac{\alpha_0}{z^4} \equiv -\frac{C_4}{z^4}, \qquad (2.57)$$

where the DC polarizability α_0 is sufficient to describe the properties of the atom. Lifshitz generalized the treatment of surface forces, yielding the London and Casimir-Polder regimes as limiting cases. The theory also includes the generalization to dielectric media, wich leads to a weaker potential

$$U_{\text{CP}}(z) = -\frac{3\hbar c}{32\pi \epsilon_0} \frac{\alpha_0}{z^4} \frac{\epsilon_r - 1}{\epsilon_r + 1} \phi(\epsilon_r) \equiv -\frac{C_{4,d}}{z^4} \qquad (2.58)$$

2.3 Surface forces

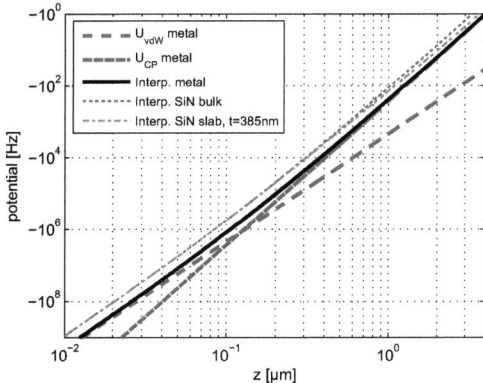

Figure 2.7.: Surface potential of a perfect conductor and a dielectric bulk or slab SiN. Shown are the asymptotic potentials in the van der Waals-London and Casimir-Polder regimes as well as the potential of a thin dielectric slab.

with a numerical factor $\phi(\epsilon_r)$ which is 0.81 (0.46) for Si (SiN) having a dielectric constant $\epsilon_r = 13.69$ (4.08) [119].

The short range van der Waals-London and the retarded Casimir-Polder regime can be summarized by an interpolation formula that describes a smooth transition between the two regimes,

$$U_{\mathrm{CP}}(z) = -\frac{C_4}{z^3(z - \lambda/2\pi)}, \tag{2.59}$$

where the distance of the crossover is given by the reduced wavelength of the strongest atomic dipole transition ($\lambda/2\pi \approx 120$ nm for ^{87}Rb). This implies that the two strength coefficients are connected via $C_4/C_3 = \lambda/2\pi$. Figure 2.7 shows the two asymptotic potentials and the interpolation Eq. 2.59.

The interpretation of the various forces is an interesting question for itself. Especially, whether the CP potential could be regarded as a retarded van der Waals-London force was discussed [121, 122]. It was found that in the CP regime, the physical origin has to be described as a distance dependent AC Stark shift of the atomic ground state which is caused by quantum fluctuations of the vacuum, rather than by retarded image fields.

Thermal (Lifshitz) regime

In the above discussion, the CP force originates from zeropoint fluctuations of the vacuum. The situation is different when thermal populations of the vacuum modes are present. Thermally excited modes will have a wavelength $\lambda > \lambda_T = \hbar c/k_B T$, and the surface potential will differ from U_{CP} for distances large compared to the thermal wavelength λ_T. In a simple form [123], the resulting potential for a dielectric can be written as

$$U_{\text{th}}(z) = -\frac{1}{16\pi\epsilon_0} \frac{\alpha_0 k_B T}{z^3} \frac{\epsilon_r - 1}{\epsilon_r + 1}, \tag{2.60}$$

which is now dependent on the temperature T of the surrounding environment defining the temperature of the vacuum, and Planck's constant is no longer contained.

Thin layers

So far, the formulas are derived by assuming the dielectric as a semi-infinite, homogeneous medium. The scaling changes when the thickness h of the dielectric medium is much less than the atom-surface distance d, $h \ll d$. Calculations show that in this limit, the short range van der Waals-London and the intermediate range Casimir-Polder forces scale as [124, 125, 126]

$$U_{\text{vdW}}(z,h) \propto -\frac{C_3 h}{z^4} \tag{2.61}$$

$$U_{\text{CP}}(z,h) \propto -\frac{C_{4,d} h}{z^5}. \tag{2.62}$$

For conductors the behaviour is different, and the penetration depth becomes an important parameter. As a general result one can state that thin metal layers have essentially the bulk surface potential as soon as the reflectivity comes close to the bulk value. E.g. for Au this occurs for a film thickness $h \gtrsim 30$ nm.

When several thin layers are combined, the non-additivity of surface forces has to be accounted for. In [127], the van der Waals potential of a dielectric waveguide consisting of a several stacked thin films is calculated. The potential for an interface between vacuum with refractive index $n_1 = 1$ and a stack of two dielectric layers with refractive index n_2, n_3 is found to be

$$U_{\text{vdW}}(z,t) = -C_3 \left(\frac{\alpha_{21}}{z^3} + \frac{\alpha_{21}^2 - 1}{\alpha_{21}} \sum_n^\infty \frac{(\alpha_{21}\alpha_{23})^n}{(z+nh)^3} \right), \tag{2.63}$$

where $\alpha_{ij} = (\epsilon_{r,i} - \epsilon_{r,j})/(\epsilon_{r,i} + \epsilon_{r,j})$. The layers lead to an enhancement of the potential due to multiple reflections of the electromagnetic waves inside the waveguide layer.

Finally we want to mention the possibility of repulsive CP forces by either realizing thermal non-equilibrium situations [123] or by using magneto-dielectric materials [125].

2.3 Surface forces

Measurements of van der Waals - London and Casimir-Polder forces

Measurements of fundamental surface forces has been a long standing topic. Using atoms as a sensitive probe has proven to be a powerful method to investigate this physics, especially at large atom-surface distance. Experiments include transmission measurements with Rydberg atoms [128, 129], reflection of ultracold atoms from evanescent waves [130, 131], examination of surface induced atom loss [24], precision measurements of trap frequency shifts [26, 27], or quantum reflection from the surface [132, 133, 134].

2.3.2. Adsorbates and stray charges

The microscopic state of a surface is a dirty affair, dominated by a high concentration of defect states, electric and magnetic impurities, oxide layers, and adsorbed atoms or molecules which can be the origin of electric or magnetic stray fields. Ultracold atoms trapped close to the surface are sensitive to electromagnetic fields and can thus be affected by the fields emanating from the surface.

Stray field potentials from adsorbates

One particular origin of surface potentials are stray fields from alkali adsorbates that are deposited during the course of experiments with ultracold atoms close to a surface. We review this effect in more detail since it is a likely origin for the surface potential that we observe in our experiments.

For distances of the order of the Bohr radius, the attractive van der Waals-London force is overwhelmed by a repulsive force originating from the Pauli exclusion principle. An atom accelerated towards a surface should be reflected at the repulsive potential when no dissipative process occurs. However, one observes that a large fraction of atoms sticks on the surface and forms adsorbates. The inelastic process associated with this behaviour can be described in the language of Feshbach resonances [135]. The atom is deexcited to a local bound state, where the energy difference is resonantly coupled to a phonon excitation of the local surface configuration. Due to the high density of states of surface excitations, inelastic atom-surface collisions are likely and the probability to stick at the surface is usually high.

Alkali atoms adsorbed on a metal surface suffer from level broadening and a deformation of the lowest S and P orbitals of the valence electron due to interaction with the energy bands of the substrate. If the work function ϕ_s of the substrate is larger than the ionization energy E_i of the atom, the valence electron will be transferred partially to the substrate, and a dipole moment μ_{el} aligned normal to the surface is formed [136, 137, 138, 139, 25]. Figure 2.8 (a) depicts the situation. The resulting stray field of the dipoles of a collection of adatoms can become sizable and lead to a polarization of trapped atoms nearby. When the adatoms are deposited in a non-uniform density distribution, the electric stray field can have strong gradients and lead to an attractive potential

$$U_{\text{ad}}(z) = -\frac{\alpha_0}{2}|E(z)|^2. \tag{2.64}$$

Depending on the density distribution and the number of adatoms, the potential can be larger than the CP potential and dominate the atom-surface interaction.

The distance-dependence of U_{ad} depends on the spatial distribution of adsorbates on the surface. E.g. a line of dipoles would cause a potential $\sim 1/z^4$, while for a point-like distribution it would fall off as $\sim 1/z^6$. In cold atom experiments, strongly inhomogeneous distributions can originate from the deposition of BECs on the surface. The condensates from a cigar-shaped trap would result in an elongated distribution of adsorbates. To estimate the stray field of such a distribution, we assume a gaussian 2D density distribution

$$n(x,y) = n_0 \exp\left(-\frac{x^2}{(2R_{\text{TF},x})^2} - \frac{y^2}{(2R_{\text{TF},y})^2}\right) \tag{2.65}$$

where the choice of $2R_{\text{TF},i}$ as radius accounts for some diffusion on the surface. The number of adsorbed atoms is given by $N_{ad} = \iint n(x,y)dxdy$. The electric field $E(z)$ is obtained by an integration over the density distribution of the adatoms

$$E(z) = \iint dx'dy' \frac{en(x',y')}{4\pi\epsilon_0}\left[\frac{1}{(z-\frac{\mu_{el}}{2e})^2 + x'^2 + y'^2} - \frac{1}{(z+\frac{\mu_{el}}{2e})^2 + x'^2 + y'^2}\right] \tag{2.66}$$

with e the electron charge, and where the individual positive and negative charge contributions of the dipole are summed explicitly.

Figure 2.8 (b+c) shows a typical adsorbate density distribution for deposited BECs and the resulting potential $U_{\text{ad}}(z)$. As an upper limit for the inhomogeneity of the adsorbate distribution we calculate the density distribution with Eq. 2.65 for a BEC of $N_c = 2000$ atoms in a trap with $\omega_{x,y} = 2\pi \times [1, 10]$ kHz. The stray field potential is shown for $[100, 500, 2500]$ BECs deposited at the same position. We assume a dipole moment of $\mu_{el} = 3$ Debye for Rb adsorbed on Au, which has a work function $\phi_s = 4.8 - 5.1$ eV[1]. For $N_{ad} > 6 \times 10^5$ the adsorbate potential becomes larger than the CP potential. The maximum density for the largest atom number is ~ 0.5 nm^{-2} and thus only a few percent of a monolayer.

In [25, 27] the electric stray field of controllably deposited ^{87}Rb BECs was measured for various different materials. The electric dipole moment of Rb on Si and Ti (work function $\phi_s = 4.8$ eV and 4.3 eV respectively, to be compared to the ^{87}Rb

---

[1] We have found no value for the dipole moment of Rb adsorbates on evaporated Au in the literature. The choice of the value is motivated by the similar work function for Si and the measured μ_{el} in [25].

2.3 Surface forces 31

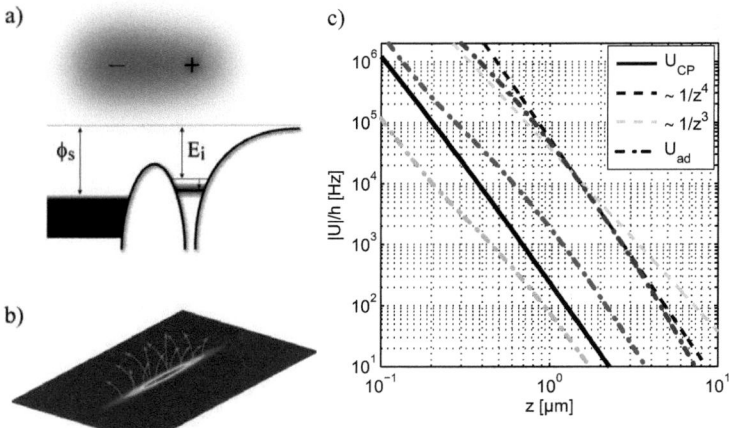

Figure 2.8.: (a) Potential energy diagram for an alkali adatom adsorbed to a metal surface. The adatom electronic ground state is shifted and broadened. For Rb on Au, the ionization energy E_i of Rb is smaller than the work function ϕ_s of gold, leading to a partial charge transfer to the surface and thus a dipole moment. (b) Elongated gaussian density distribution of adsorbed Rb atoms as expected from a deposition of BECs on the surface. (c) Adsorbate potential for a distribution as in (b) with $R_{\text{TF},i} = [3, 0.3]$ μm. Adsorbate potentials are shown for $N_{ad} = (2 \times 10^5, 1 \times 10^6, 5 \times 10^6)$ atoms (from light to dark grey). For comparison, the CP potential and potentials of the form A_δ/z^δ with $\delta = (3, 4)$ fit to the potential for $N_{ad} = 5 \times 10^6$ are shown.

ionization energy of $E_i = 4.2$ eV) was found to be $\mu_{el} \sim 3 - 15$ Debye [25], while fused silica with no expected charge transfer also showed ~ 3 Debye [27].

The surface coverage of adsorbates Θ plays an important role for the strength of the dipole. One effect is that the work function of the surface approaches the value of the metallic adsorbate species already for small coverage $\Theta_0 \sim 0.3$ [137, 138]. Yet this does not imply that no dipole moment remains for $\Theta \geq \Theta_0$, as the deformation of the lowest orbitals also contributes to polarization. For high coverage one expects a residual dipole comparable to the value for a dielectric surface.

A related question is the steady state coverage of the surface depending on the partial pressure of Rb atoms. For a given partial pressure p, the adsorption rate for atoms of mass m and temperature T_a is given by the flux of atoms impinging on the surface

$$r_{ad} = S(\Theta) \frac{p}{\sqrt{2\pi m k_B T_a}}, \tag{2.67}$$

which depends on the surface coverage via the sticking probability $S(\Theta)$. On the other hand, the desorption rate is given by an Arrhenius law

$$r_{de} = \nu_0(\Theta) N_0 \Theta \exp\left(-\frac{E_b(\Theta)}{k_B T_s}\right) \qquad (2.68)$$

with the surface temperature T_s, the eigenfrequency of the bound state ν_0 (typically of the order of 10^{13} Hz), the binding energy E_b, and the number of adsorbed atoms $N_{ad} = \Theta N_0$ being a fraction of a monolayer with N_0 atoms per unit area. The balance between these two rates determines the dynamics and the steady state value of the surface coverage,

$$N_0 \frac{d\Theta}{dt} = r_{ad} - r_{de}. \qquad (2.69)$$

To obtain meaningful predictions from this rate equation, adequate models or measured coverage dependences for $S(\Theta), \nu(\Theta), E(\Theta)$ are required. In the founding theory of Langmuir [136, 140] one assumes $S(\Theta) = S_0(1 - \Theta)$ with the zero coverage sticking probablilty S_0 often set to unity. This simple form describes a very idealized situation of a uniform surface with only one adsorption process, no interaction between adsorbates, and the formation of one single layer. Extensions of the Langmuir theory to multilayer adsorption in the BET theory [141] or the inclusion of interactions between adsorbed and gaseous adsorbates [142] proved to be more successful to describe experiments.

In the limit where $r_{ad} \gg r_{de}$, desorption can be neglected and Eqs. 2.69 with 2.67 result in an exponential law $\Theta(t) = 1 - \exp(-t/\tau_m)$, where τ_m is the time constant for the formation of a monolayer,

$$\tau_m \approx \frac{N_0 \sqrt{2\pi m k_B T_a}}{p} = 6 \times 10^{-5} \text{ s} \times \frac{1}{p \text{ [mbar]}}. \qquad (2.70)$$

For the case of Rb thermalized at $T_a = 300$ K with a partial pressure of 10^{-10} mbar it is about ten days. Figure 2.9 shows a numerical integration of Eq. 2.69 for an initially uncoated Au surface exposed to Rb with partial pressure p. For comparison, the exponential law with the time constant from Eq. 2.70 is shown.

For Rb, the condensed metal has a work function $\phi_s = 2.2$ eV which is comparable or even larger than the binding energy of Rb adsorbed to Au. This makes the formation of multilayers favourable.

Desorption is not the only process to reduce the surface coverage, e.g. Rb can form an alloy with Au, which also changes the work function of the surface. An other process for alkali atoms adsorbed on metals with large work function (as for Rb on Au) is that the atom gives its valence electron to the surface and desorbs as an ion.

2.3 Surface forces

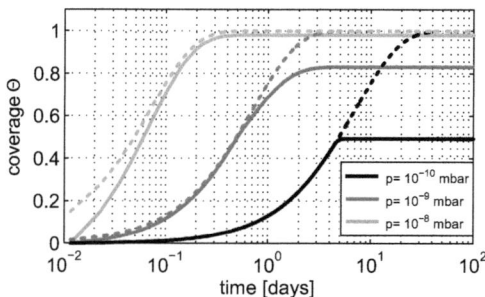

Figure 2.9.: Numerical integration of the adsorption dynamics according to Eq. 2.69 (solid lines) and exponential growth (dashed lines) valid for the limit $r_{ad} \gg r_{de}$. For the numerical calculation we use the Langmuir sticking coefficient with $S_0 = 1$, estimate the binding energy following measurements of Cs and K on Ni [137] (decreasing from $E_b(0) = 2.2$ eV to $E_b(0.5) = 1.1$ eV and constant for larger Θ), and use $\nu_0 = 10^{13}$. The temperatures are chosen to be $T_a = T_s = 300$ K. The equilibrium coverage does not reach unity due to the decrease of E_b with Θ.

The binding energy of alkali adatoms on metal is of the order of $0.1 - 3$ eV and can thus be comparable to the room temperature energy $k_B T = 26$ meV. Weak binding leads to thermal diffusion of adatoms across the surface and enables fast desorption. In [27] the diffusion of controllably deposited adatoms was determined by measurements of the decay time of the stray field. It showed to be similar for the studied Yttrium and fused silica surfaces and amounted to 8 and 2 days respectivley at room temperature. The data allowed to extract the activation energy 0.4 eV and the frequency $\nu_0 = 10^2$ Hz, the latter being ~ 10 orders of magnitude smaller than the expected eigenfrequency of the bound state.

In summary, already small amounts of adsorbates can play an important role for cold atom experiments close to surfaces. However, the situation is rather unclear for the case of high coverage. At the relatively high partial pressures used in our experiments, the surfaces will be coated with a first layer with coverage of $\sim 60 - 90$ % and several more layers that may form islands. The stray field of additionally deposited adsorbates will be reduced with respect to a deposition on an uncoated surface, and diffusion will be faster. But there are also effects that increase stray field potentials: E.g. a homogeneous field of a uniform adsorbate distribution also polarizes atoms nearby and can thereby increase the effect of an additional non-uniform distribution. Quantitative estimates of the strength of controllably deposited adsorbates in this regime are thus difficult.

Patch potentials and stray charges

Small-scale random fluctuations in the potential of conducting materials associated with strain, irregularities, or impurities, lead to electrostatic potentials and thus additional attractive forces on nearby atoms. Typical patch potentials are of the order of 100 mV with a gaussian patch size distribution with rms diameter of $D \sim 50$ nm [129, 143]. The field strength at the surface is a few hundred V/cm, but for distances $d \gg D$ from the surface the potentials average out and the remaining gradient is small.

A different source of electric stray fields can be stray charges on dielectrics. The field of only 6 elementary charges located in a region with diameter $D \ll d$ causes an attractive potential of the same magnitude as the CP potential. The shape of the potential is $|E|^2 \propto 1/z^4$ and thus identical to the CP potential in the retarded regime. The influence from stray charges can be important e.g. for laser irradiated surfaces, where ionization is expected. It was recently found that this has strong impact for ion traps with laser irradiated dielectrics close to the trap [144]. But also the ionization of alkali adsorbates on metal surfaces can create ions on neighbouring dielectric surfaces.

Magnetic impurities

Impurities with magnetic moments embedded in the substrate close to the surface can affect the trapping potential. Due to the vector nature of the magnetic field and the structure of magnetic traps, a stray field will (except for very few special cases) affect axial and radial trapping frequencies and the center of the trap simultaneously. As an example we sketch two worst case situations.

First we consider the change in trap frequency when the magnetic stray field contains a component parallel to the trap axis B_x^*. The strongest effect is on the transverse trap frequencies $\omega_\perp \propto 1/\sqrt{B_0}$ (see equation 2.11). The relative change $\delta\omega_\perp/\omega_\perp = B_x^*/(2B_0^{3/2})$ can be significant for small B_0. E.g. for a typical field $B_0 = 1$ G, a field of $B_x^* = 5$ mG changes $\omega_\perp$ by one percent. At a distance $d = 1.5$ μm from the surface, such a field could be generated e.g. by a magnetic particle (e.g. Co) of ~ 20 nm diameter magnetized along the x−axis, corresponding to $\sim 10^7$ atoms.

The effect of magnetic stray fields on the trap position is largest when the field is perpendicular to the trap axis. A calculation for the geometry and the trap parameters as used in our experiments predicts that a position shift by 10% of the ground state size in the trap requires a particle at least ten times as large as in the previous example.

Gradients of the stray field add a position dependence to both examples and additionally introduce forces.

2.4. Effects of surface forces on atom trapping

In this section I describe the effects of surface forces on ultracold atoms trapped at small distance from the surface. The static deformation of the trapping potential is analyzed and the arising loss mechanisms are introduced. Trap deformation is the basis for the dynamical coupling of ultracold atoms to mechanical motion, which is discussed in chapter 5.

2.4.1. Potential deformation

At small atom-surface distance, the magnetic potential U_m used to trap the atoms is substantially modified by the surface potential $U_s = U_{\text{CP}} + U_{\text{ad}}$, where U_{ad} includes all the additional potentials that can arise (see above). In the direction perpendicular to the surface, the combined potential is (see Fig. 2.10)

$$\begin{aligned} U[z] &= U_m + U_{\text{CP}} + U_{\text{ad}} + U_{\text{grav}} \\ &= \frac{1}{2} m \omega_{z,0}^2 (z - z_{t,0})^2 - \frac{C_4}{(z - z_c)^4} + U_{\text{ad}}[z - z_c] + mgz. \end{aligned} \quad (2.71)$$

Here, z_c is the position of the cantilever surface, C_4 the CP-coefficient, $z_{t,0}$ the minimum position of the magnetic trap, $\omega_{z,0}$ the trap frequency of the trap far away from the surface, m the atomic mass, and g the acceleration in the gravitational potential U_{grav}.

Trap depth reduction The most obvious effect of the surface potential U_s is to reduce the trap depth from the value of the magnetic potential at the surface $U_m[z_c - z_{t,0}]$ to U_0 (see Fig. 2.10) [24]. This results in a smooth barrier.

Frequency shift The curvature of U_s gives rise to a shift of the trap frequency [119] from $\omega_{z,0}$ in the unperturbed magnetic trap to

$$\omega_z^2 = \omega_{z,0}^2 + \frac{1}{m} \int n_0(z) \frac{\partial^2 U_s}{\partial z^2} dz, \quad (2.72)$$

where the normalized 1D column density $n_0(z)$ along the z-axis enters to account for the variation of the curvature across the extension of the atom cloud. For a BEC in the Thomas-Fermi regime one finds

$$n_0(z) = \frac{15}{16} \frac{1}{R_{\text{TF},z}} \left(1 - \frac{(z - z_t)^2}{R_{\text{TF},z}^2} \right)^2. \quad (2.73)$$

For a surface potential of the form of U_{CP}, the integral in Eq. 2.72 can be carried out analytically and one obtains

$$\omega_z^2 = \omega_{z,0}^2 - \frac{20 C_4}{m d^6} \frac{1}{(1 - (R_{\text{TF},z}/d)^2)^3}. \quad (2.74)$$

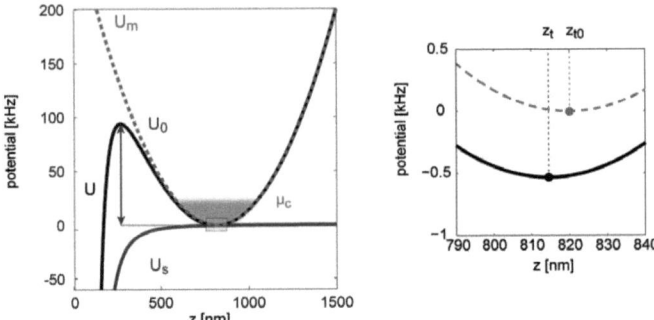

Figure 2.10.: Calculation of the deformation of the magnetic trapping potential by the Casimir-Polder potential for small atom-surface distance. It leads to a barrier of height U_0, a shift of the trap minimum and a change in trap frequency. Parameters are $\omega_z/2\pi = 10$ kHz, $d = 820$ nm, $U_s = U_{\rm CP}$. The chemical potential μ_c of a BEC with $N = 600$ atoms is shown for comparison. Right panel: Enlarged view of the potential close to the trap minimum (indicated by the grey box in the left panel). The shift of the trap minimum is visible.

When the distance to the surface $d = z_t - z_c$ is much larger than the cloud radius, the last factor in Eq. 2.74 is close to unity and it is a good approximation to neglect the spatial extent of the cloud. For an arbitrary surface potential, the trap frequency is then given by

$$\omega_z^2 = \omega_{z,0}^2 + \frac{1}{m}\frac{\partial^2 U_s}{\partial z^2}, \tag{2.75}$$

where the derivative of U_s is taken at the c.o.m. position.

Trap anharmonicity The surface potential introduces anharmonicity to the trapping potential. This results in a dependence of the oscillation period $\mathcal{T}$ on the oscillation amplitude. In general, the oscillation will be asymmetric, and the classical turning points (z_-, z_+) which define the oscillation amplitude $b = (z_+ + z_-)/2$ can be found via the relation $U(z_\pm) = E$, where $E = m\dot{z}^2/2 + U(z)$ is the total particle energy. The oscillation period in an arbitrary potential can be written as

$$\mathcal{T} = \sqrt{2m} \int_{z_-}^{z_+} \frac{dz}{\sqrt{E - U(z)}}, \tag{2.76}$$

which directly links to the effective frequency of the oscillation $\omega_{\rm com} = 2\pi/\mathcal{T}$. However, the singularities at (z_-, z_+) detain from an evaluation of the integral unless $U(z)$ leads to an elliptical form of the integrand. Various techniques are proposed

2.4 Effects of surface forces on atom trapping

to approximate $\mathcal{T}$ by proper expansions of the integrand [145, 146]. For small amplitudes, ω_{com} can be derived from an expansion of the potential around the trap minimum position [119], yielding

$$\omega_{\text{com}}^2(b) \approx \omega_z^2 + \frac{b^2}{8m}\frac{\partial^4 U_s(z_t)}{\partial z^4}. \tag{2.77}$$

For traps close to a surface, the second term in Eq. 2.77 can be sizable and introduce a significant amplitude dependence of the oscillation frequency. In chapter 5.8 we show that anharmonicity causes dephasing of collective oscillations and is a limiting factor for the sensitivity of coupling measurements.

Minimum shift Finally, the gradient of U_s leads to a shift of the center of mass position from $z_{t,0}$ to

$$z_t = z_{t,0} - \int n_0(z) \frac{1}{m\omega_z^2}\frac{\partial U_s}{\partial z}dz \approx z_{t,0} - \frac{1}{m\omega_z^2}\frac{\partial U_s}{\partial z}. \tag{2.78}$$

The approximation is again valid for $d \gg R_{\text{TF},z}$ and in this case coincides with the new trap minimum position z_t. The right panel of Fig. 2.10 shows the shift of the trap position.

The deformation of the trapping potential can e.g. be used to analyze the surface potential. In the group of Eric Cornell, the change of the trap frequency was used to measure the strength of the Casimir-Polder potential [26] or the shape and strength of adsorbate potentials [25, 27].

In our experiments, we use the dynamic modulation of the trap deformation to couple mechanical motion to collective motion of the atoms in the trap.

2.4.2. Sudden loss and surface evaporation

When a thermal or partially condensed cloud is quickly brought close to a surface where the trap depth U_0 is reduced to a value of the order $\lesssim 8k_BT$, a notable fraction of the atoms has higher energy than the barrier and will be lost from the trap within one oscillation period, so called "sudden loss". The trap will be modified mainly along one axis by the surface potential, and the situation is described by a 1D geometry. The 1D sudden loss of a thermal cloud corresponds to a truncation of the 1D Boltzmann distribution at the energy U_0, leaving a fraction of remaining atoms [24]

$$\chi = 1 - e^{-\eta}, \tag{2.79}$$

where $\eta = U_0/k_BT$ is the ratio between the remaining trap depth and the thermal energy. The situation for a partially condensed cloud is more complicated since the mean field repulsion affects the energy distribution of the thermal component. We discuss this in section 5.3.

Surface evaporation leads to additional atom loss on the timescale of typical trap lifetimes. The evaporation rate equals the rate at which atoms with energy larger than U_0 are reproduced by elastic collisions. It is quantified by [147, 148, 24]

$$\Gamma(\eta) = f(\eta) e^{-\eta}/\tau_{el}. \tag{2.80}$$

Elastic collisions occur at a rate $\tau_{el}^{-1} = \sqrt{2}\langle n \rangle \sigma_{el} \bar{v}$, given by the mean density $\langle n \rangle$, the average relative velocity between two atoms $\sqrt{2}\bar{v}$, and the elastic scattering cross section σ_{el}. For a condensate, $\bar{v}_c$ has to be evaluated via the kinetic energy, $\bar{v}_c = \sqrt{2E_{\mathrm{kin}}/m}$, while for a thermal cloud $\bar{v}_{th} = \sqrt{8k_B T/\pi m}$. The scattering cross sections are $\sigma_{el,c} = 4\pi a_s$ and $\sigma_{el,th} = 8\pi a_s$ respectively, and the mean densities $\langle n_c \rangle = 4\mu_c/7g$ and $\langle n_{th} \rangle = N/(2\pi k_B T/m\omega_{ho}^2)^{3/2}$ (note that the condensate and especially the thermal mean density is only a rough estimate, a more accurate determination has to follow Eqns. 2.46 and 2.47).

The geometry of the situation enters via the dimensionless factor

$$f(\eta) = 2^{-5/2}(1 - \eta^{-1} + \frac{3}{2}\eta^{-2}), \tag{2.81}$$

which accurately describes 1D evaporation for $\eta \geq 4$ and which is roughly a factor 4η smaller than in the case of 3D evaporation [148, 147]. Evaporation is only important for times $t \gg \tau_{el}$. Figure 2.11 in chapter 2.5.2 shows the lifetime limitation due to surface evaporation for a thermal cloud. It is important for distances $d < 10$ μm.

As the lost atoms carry away more than the average energy per atom, evaporation also leads to cooling. In [149, 150] this effect was used for the preparation of BECs.

2.4.3. Tunneling

We now discuss effects that affect mainly the condensate. For traps in extreme vincinity to the surface, the potential barrier between the surface and the trap can be of the same size as the atomic energy and extend over a width smaller than e.g. the size of the ground state wave function, such that tunneling through the barrier becomes an important process. The transmission coefficient T of the barrier can be calculated in the WKB approximation [151, 152, 153, 154]

$$T(E, U) = \exp\left\{-2 \int_{x_1}^{x_2} \sqrt{\frac{2m}{\hbar^2}(U(z) - E)} dz\right\} \tag{2.82}$$

where E is the expectation value of the energy of an atom and $x_{1,2}$ are the classical turning points of the atom in the potential $U(z)$. The tunneling rate is then given by

$$\Lambda = \frac{\omega_z}{2\pi} T(E, U), \tag{2.83}$$

2.4 Effects of surface forces on atom trapping

where the trap frequency gives the rate of bounces from the barrier. Tunneling contributes only, when the atomic energy is close to U_0. Due to the exponential dependence on the energy difference, the effect of tunneling can be summarized by a slight reduction of the barrier height (typically a few percent, see chapter 5.3).

2.4.4. Quantum reflection

From classical wave mechanics it is known that an impedance mismatch causes reflection: A wave gets reflected, if the medium in which it propagates changes its impedance such, that the local wavelength λ is changed substantially within a distance smaller than λ. In optics this occurs for jumps in the refractive index and leads to reflection on boundaries. For matter waves propagating in an external potential, quantum reflection occurs when the local deBroglie wavelength $\lambda_{dB}(z) = \hbar/p(z)$ with the local momentum $p(z) = \sqrt{2m(E-U(z))}$ changes substantially within one wavelength. This violates the condition for the validity of the WKB approximation

$$\frac{1}{2\pi}\left|\frac{d\lambda_{dB}}{dz}\right| = \hbar\left|\frac{m}{p(z)^3}\frac{dU(z)}{dz}\right| \ll 1. \qquad (2.84)$$

When $d\lambda/dz$ is markedly nonzero, the amplitude of the reflected wave becomes significant, and the respective regions are called "badlands".[2]

Purely attractive potentials can also show quantum reflection. When analyzing the maximum of WKB violation in Eq. 2.84, one finds that reflection occurs at a position, where the modulus of the potential is of the same magnitude as the incident energy [155, 135]. For potentials of the form $U_\alpha(z) = -C_\alpha/z^\alpha$ one can see that for $\alpha > 2$, quantum reflection occurs at a finite distance from the surface, e.g. for the Casimir-Polder potential $z_{\text{refl}} = (C_4/5E)^{1/4}$. In [155], approximate expressions for the reflection probability are deduced. For potentials of the form U_α, the reflection probability R is found to be

$$R_\alpha \approx \exp(-2b_\alpha k) \qquad \text{for } k \to 0 \qquad (2.85)$$

with the effective range of the potential $\beta_\alpha = \sqrt{2C_\alpha m/\hbar^2}$, the asymptotic wave vector $k = \sqrt{2mE/\hbar^2}$, and the parameter $b_{\alpha=(3,4)} = (\pi\beta_3, \beta_4)$.

Experiments with Na BECs reflected from a bulk Si surface [134] or a nanostructured Si surface [156] showed that the mean-field interaction in the cloud leads to a reduction of quantum reflection for low incident velocities. First, the repulsive mean-field interaction accelerates atoms away from the cloud center, such that the atoms aquire an average velocity $v_{\text{rep}} = \sqrt{g\langle n_c\rangle/m}$, corresponding to the speed of sound in

---
[2]Note that the given condition is neither a necessary nor a sufficient criterion for quantum reflection. A proper criterion is given in [155].

the condensate. Thus, the effective incident velocity is given by $v_{\text{eff}} = v_{\text{inc}} + v_{\text{rep}}$ and arbitrarily low values are not achievable ($\min(v_{\text{eff}}) \sim 0.2$ mm/s). Additionally, an even stronger effect was attributed to the deformation of the combined mean-field and surface potential. The spatial variation of the mean-field potential rounds off the surface potential and thus reduces reflection for low incident velocities. Despite these effects, normal incidence quantum reflection with up to 30% reflectivity for a bulk Si [134] and up to 67% for a nanostructured Si surface with a strongly reduced surface potential [156] was demonstrated.

In our experiments, BECs are prepared close to the surface with small $v_{\text{inc}} \sim 0 - 5$ mm/s, but large $v_{\text{rep}} \sim 4 - 10$ mm/s due to high trap frequencies. For an unmodified surface potential, only small reflectivity $\sim 10^{-3}$ is expected. However, in combination with a magnetic trap, quantum reflection is relevant when the combined potential shows a barrier of height $U_0 \lesssim \mu_c$. This situation is simlar to the proposals [157, 158], and for moderate ω_z an effect should be observable.

2.5. Atom loss and heating

In this section I summarize loss and heating effects that lead to a lifetime reduction independent of the modification of the trapping potential. In particular, three-body-collisional loss and technical heating impose severe limitations at the high trap frequencies that are desired for our experiment.

2.5.1. Collisional loss

In typical BEC experiments, the condensate lifetime amounts to a few seconds up to a minute. Longer lifetimes are inhibited by inelastic collisions that cause trap loss.

Three important collisional loss processes are contributing: Collisions with background gas atoms, two-body, and three-body inelastic collisions. Collisions wich involve more partners are unlikely. The time evolution of the atom number is thus governed by a rate equation with three contributions,

$$\frac{1}{N}\frac{dN}{dt} = -\gamma_{bg} - K\langle n \rangle - L\langle n^2 \rangle. \tag{2.86}$$

The background loss rate γ_{bg} is due to collisions with the room temperature background gas in the vacuum chamber. Due to the high energy, collisions with large momentum transfer lead to loss of trapped atoms. The loss rate is given by $\gamma_{bg} = \sigma_{bg} p \sqrt{3/k_B T m}$, where p is the background pressure and σ_{bg} the cross section for a background gas atom to eject a trapped atom [159].

The second term describes inelastic two-body collisions which are domintated by spin-exchange processes, e.g. of the form $|2, 1; 2, 1\rangle \rightarrow |2, 2; 2, 0\rangle$. The rate depends on the internal state of the collision partners and is energetically forbidden for the

2.5 Atom loss and heating

states $|2, \pm 2\rangle, |1, \pm 1\rangle$, such that it can be avoided by preparing the atoms in proper states. A much weaker process is the spin-dipole interaction, which is negligible on typical experimental timescales.

The last term in Eq. 2.86 describes loss due to inelastic three-body collisions. When three atoms collide, two of them can form a molecule, while the third is necessary for simultaneous momentum and energy conservation. The binding energy is converted into kinetic energy of the collision partners and leads to trap loss of typically both the molecule and the atom. Heating and additional loss can occur due to collisions of these energetic particles with other trapped atoms. The loss rate is

$$\gamma_{3b} = L\langle n^2 \rangle \propto \begin{cases} \omega_{ho}^{12/5} N^{4/5} & \text{BEC in TF limit} \\ \omega_{ho}^{6} N^{2} & \text{thermal cloud,} \end{cases} \quad (2.87)$$

where L has been measured for the relevant states $|1, -1\rangle, |2, 2\rangle$ to be $L = [5.8, 18] \times 10^{-30}$ cm^{6}s^{-1}, respectively [160, 161].

For a condensate with a large thermal fraction, equation 2.86 has to be modified to account for atom bunching in the thermal cloud. This leads to [161]

$$\frac{1}{N}\frac{dN}{dt} = -\gamma_{bg} - K\left[\langle n_c \rangle + 2\langle n_T \rangle\right] - L\left[\langle n_c^2 \rangle + 6\langle n_c n_T \rangle + 6\langle n_T^2 \rangle\right], \quad (2.88)$$

which requires the knowledge of the density distributions given in Eq. 2.46.

Especially for high trap frequency (> 5 kHz), three-body collisional loss is a severely limiting process that reduces the trap lifetime to a few ms also for small clouds.

2.5.2. Thermal magnetic near-field noise

The thermal motion of electrons in a conductor at finite temperature leads to current fluctuations called Johnson-Nyquist noise. The currents cause fluctuating magnetic fields that couple to the magnetic moment of atoms. Close to the surface, the fluctuating fields are strongly enhanced compared to the thermal blackbody field far away due to the contribution of evanescent waves. This near-field regime manifests in the low frequency spectrum, where the associated wavelengths are large compared to the atom-surface distance. The frequency components of the fluctuations at the Larmor frequency of the atoms in the trap can drive spin-flip transitions, leading to trap loss and decoherence. The effect was predicted by Henkel $et\ al.$ [162, 163] and demonstrated experimentally in the groups of Hinds [28], Cornell [29] and Vuletic [24].

The spectral density of magnetic field fluctuations at a distance d from a solid with conductivity σ at temperature T is given by [163]

$$S_B(\omega) = \frac{\mu_0^2 \sigma k_B T}{16\pi d} s_{\alpha\beta} g(d, h, w, \delta), \quad (2.89)$$

with $s_{\alpha\beta} = \text{diag}(\frac{1}{2}, \frac{1}{2}, 1)$ a diagonal tensor with the z-axis normal to the surface. The dimensionless function $g(d, h, w, \delta)$ depends on the thickness h and width w of the conductor, and on the skin depth $\delta = \sqrt{2/\sigma\mu_0\omega}$ being the only frequency dependent factor. For typical Larmor frequencies of about one MHz and gold metallization, δ is of the order of 100 µm and $\delta \gg (d, h, w)$ for typical wires. In the near field ($z \ll c/\omega$), analytic approximations for g exist [162, 164]

$$g = \begin{cases} \left(1 + \frac{2d^3}{3\delta^3}\right)^{-1} & \text{metallic half space,} \\ \frac{h}{h+d} \cdot \frac{w}{w+2d} & \text{thin wire}, \delta \gg (h, d). \end{cases} \qquad (2.90)$$

To minimize magnetic near field noise it is thus beneficial to minimize the amount of metallization close to the atoms. Cooling the substrate to reduce thermal fluctuations will in general not be helpful since the conductivity increases for lower T and the product $\sigma(T)T$ even increases for decreasing T. However, the situation changes in the superconducting regime, where magnetic noise is reduced by orders of magnitude [165, 166, 167].

The spin-flip rate between two states $|i\rangle, |f\rangle$ caused by the magnetic field noise can be calculated by Fermis golden rule [163]

$$\gamma_s = \sum_{\alpha=x,y,z} \frac{|\langle i|\mu_\alpha|f\rangle|^2}{\hbar^2} S_{B\alpha\alpha}(\omega_{fi}). \qquad (2.91)$$

For atoms trapped in the state $|i\rangle = |2, 2\rangle$, trap loss occurs after the cascade process $|2, 2\rangle \to |2, 1\rangle \to |2, 0\rangle$ [24] with the individual transition matrix elemtents $\mathcal{M}_{12,yz} = \langle 1|\mu_{y,z}|2\rangle = \mu_B/2$, $\mathcal{M}_{01,yz} = \langle 0|\mu_{y,z}|1\rangle = \sqrt{3/8}\mu_B$, and $\mathcal{M}_{12,x} = \mathcal{M}_{01,x} = 0$ where the spin is assumed to be oriented along x. Summed up, the factor for the matrix elements is $\mathcal{M}_{02,yz}^2 = (\mathcal{M}_{12,yz}^{-2} + \mathcal{M}_{01,yz}^{-2})^{-1} = 3/20$ and the spin-flip rate above a thin wire becomes

$$\gamma_s = \frac{9\mu_0^2\mu_B^2\sigma k_B T}{640\pi\hbar^2 d} \frac{h}{h+d} \frac{w}{w+2d}. \qquad (2.92)$$

Figure 2.11 shows a calculation of the spin-flip rate together with surface evaporation according to Eq. 2.80, three body collisional loss according to Eq. 2.87, and background loss for a thermal cloud above a room temperature trapping wire.

Magnetic noise can also drive transitions between different hyperfine states $|2, i\rangle \leftrightarrow |1, f\rangle$ with a transition frequency $\omega_{fi} = 2\pi \times 6.8$ GHz. At such high frequencies, $\delta = 0.9$ µm and the thin wire limit $\delta \gg h$ is not valid for typical geometries. The dependence of g on δ results in smaller noise spectral density, and the trap lifetime will not be limited by these transitions.

While the components of the field noise transverse to the atomic spin $S_\perp(\omega)$ can drive spin-flips, parallel components $S_\parallel(\omega)$ contribute to randomization of the phase

2.5 Atom loss and heating

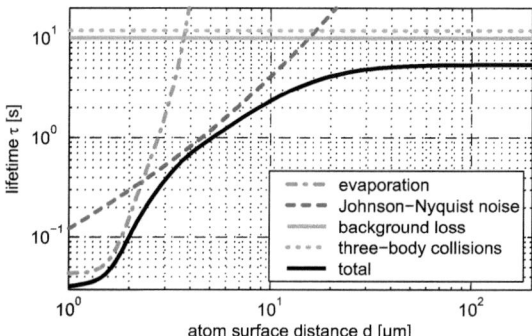

Figure 2.11.: Trap lifetime above a room temperature gold wire ($w = 50$ μm, $h = 4$ μm) for 3000 atoms in a trap with $(\omega_\perp, \omega_x) = 2\pi \times (1000, 100)$ Hz at $T = 1.5\, T_c$. For large distance, background and three-body collisional loss limit the lifetime. For intermediate distance, magnetic near-field noise induced spin-flip loss is the limiting process. Very close to the surface, the reduced trap depth leads to fast evaporation. The kink in the evaporation lifetime occurs where the trap depth becomes to low for the 1D evaporation model to be valid.

of a state and lead to dephasing e.g. of spin superposition states.

At low frequencies, spatially inhomogeneous field fluctuations contribute to heating and decoherence of atomic c.o.m. motion [163]. The length scale for the inhomogeneities is of the order of the distance d between the trap and the metallization. From an evaluation of Fermi's golden rule now for spatial wave functions $\phi(\boldsymbol{r})$, one obtains a transition rate from the ground to the first excited harmonic oscillator state of

$$\gamma_h \approx \frac{\mu_\parallel^2}{\hbar^2} \frac{a_{\mathrm{ho}}^2}{d^2} S_\parallel(\omega_t) \approx \frac{a_{\mathrm{ho}}^2}{d^2} \gamma_s, \tag{2.93}$$

where the parallel component of the magnetic field noise at the trap frequency ω_t has to be evaluated and the size of the wave function a_{ho} enters. Heating is thus reduced by a factor $(a_{\mathrm{ho}}/d)^2$ compared to spin-flips and will be negligible unless the cloud approaches the surface to $d \approx a_{\mathrm{ho}}$.

Spatially homogeneous fluctuations can lead to jitter of the trap position by adding to the DC fields that create the trap. The associated heating rate can be estimated by [82]

$$\gamma_h \approx \frac{\omega_t}{\omega_L} \gamma_s. \tag{2.94}$$

As $\omega_t \ll \omega_L$ in typical traps to avoid Majorana loss, this rate is small compared to γ_s. Note that fluctuations leading to a jitter of the trap curvature at frequency $\omega = 2\omega_t$ excite the parametric resonance and also contribute to heating, however at a rate substantially smaller than the other given rates.

2.5.3. Technical heating

Technical fluctuations of the magnetic trapping potential lead to heating and decoherence of a BEC. As the heating rates scale strongly with the trap frequency, tight traps require exceptional low noise current sources.

Trap shaking Technical fluctuations of the trap minimum position give rise to a time dependent potential

$$U_{\text{ext}}(t) = \frac{1}{2}m\omega_z^2(z - \delta z(t))^2 \tag{2.95}$$

with fluctuations $\delta z(t)$ that are characterized by the spectral density

$$S_z(\omega) = \int_{-\infty}^{\infty} d\tau e^{-i\omega\tau} \langle \delta z(t+\tau)\delta z(t) \rangle. \tag{2.96}$$

Trap shaking leads to resonant response only for the spectral components at the trap frequencies. Small, temporally uncorrelated fluctuations that change phase and amplitude ramdomly excite the c.o.m. mode of a trapped cloud into a thermal state (see also chapter 3.1.2). This results in a temperature increase that can be calculated perturbatively [168, 169]. One obtains a heating rate

$$\dot{T} = \frac{m}{2k_B}\omega_z^4 S_z(\omega_z) \tag{2.97}$$

which depends on the spectral density at the trap frequency, scales with the fourth power of the latter, and gives rise to a linear increase in temperature. For Ioffe-Pritchard wire trap configurations (e.g. Dimple or Z-trap) with the trap axis along x and trap minimum at z_0 one finds [170]

$$S_z(\omega) = z_0^2(s_{I_x}(\omega) + s_{B_{by}}(\omega)) \tag{2.98}$$

where s_I, s_B are relative noise spectra ($s_{I,B} = S_{I,B}/(I,B)$). The magnitude of the heating rate can then be obtained by

$$\dot{T} = 3.3 \times 10^7 \frac{\text{K}}{\text{s}} \left(\frac{\omega_z}{2\pi\text{kHz}}\right)^4 \left(\frac{I_x/\text{A}}{B_{by}/\text{G}}\right)^2 \times \left(s_{I_x}(\omega) + s_{B_{by}}(\omega)\right). \tag{2.99}$$

2.5 Atom loss and heating

Fluctuations of the trap frequency Fluctuating fields also cause modulations of the trap frequency. This is well known from parametric resonances and can be modelled by a perturbation of the spring constant $\delta k = \delta(m\omega_z^2)$, and the fluctuating potential reads

$$U_{\text{ext}}(t) = \frac{1}{2}m\omega_z^2(1+\delta k)z^2(t). \tag{2.100}$$

In this case, the excitation drives transitions with $\Delta n = 2$ with n being the excitation quantum number, and the energy increase is exponentially. The resulting energy doubling rate is given by

$$\gamma_{\delta_z} = \frac{\omega_z^2}{4}S_k(2\omega_z) \tag{2.101}$$

which depends on the spectral density at $2\omega_z$ and scales quadratic with the trap frequency.

3. Micro- and nanomechanical oscillators

In this chapter, micro- and nanomechanical oscillators are introduced from the perspective of using them as coupling partners for quantum optical atomic systems. In the first section I review fundamental properties and important effects that characterize these devices. I then summarize optical techniques for the read out and manipulation of mechanical resonators, which contribute largely to the controllablilty of these devices. In the last section I turn the view to one of the fundamental goals for the research with mechanical oscillators, the control of micro- and nanomechanical motion at the quantum level and the observation of macroscopic quantum behaviour.

3.1. Fundamental properties

Micromechanical resonators are an important tool in many fields of science e.g. due to their use in atomic force microscopes (AFM) [171]. The AFM has become one of the central instruments for imaging, measuring, and manipulating matter at the nanoscale. A recent achievement is e.g. the imaging of the chemical structure of a molecule by the use of Pauli repulsion forces [172]. Apart from the broad application in AFMs, micro- and nanomechanical resonators have proven to reach extreme sensitivity in force sensing [173], mass detection [64], calorimetry [174], or electrometry [175, 176, 177, 178], down to the zepto-range (= 10^{-21}, e.g. zN). One hallmark along this line is the resolution of a single electron spin in a solid, demonstrated in the group of Dan Rugar at IBM [62], which could be improved recently to a sensitivity of ~ 10 nuclear spins by magnetic resonance force imaging (MRFM) [63]. This corresponds to a force sensitivity below one Attonewton [173]. In a different context, such high sensitivity was used to establish new constraints on non-Newtonian forces at small distances [179]. The force sensitivity is also the figure of merit for mass sensing, which ultimately heads for single atom detectivity [180, 181, 182, 64, 183]. A room temperature experiment employing a double-wall carbon nanotube recently demonstrated a sensitivity of 0.4 gold atoms $\text{Hz}^{-1/2}$ and revealed the shot noise of the atomic adsorption [64].

Beyond their importance as versatile sensors, small scale mechanical oscillators are also investigated for their own fundamental properties. Studies of classical properties like nonlinear behaviour [184, 185, 186, 187], the origin of dissipation [188, 189], and

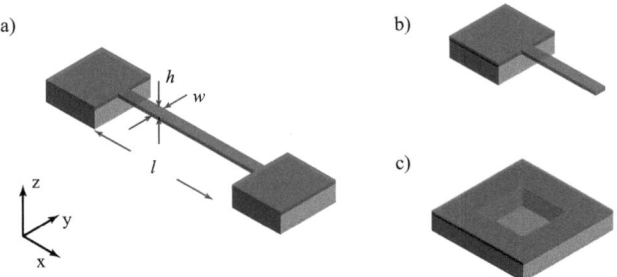

Figure 3.1.: Standard geometries of micro- and nanomechanical resonators. (a) Double clamped beam resonator, (b) single clamped beam or cantilever, (c) membrane.

noise processes in nanomechanical resonators [190] span a broad range.

Resonator Types

Micro- and nanofabrication provide techniques for the creation of a plethora of shapes for mechanical resonators on a microscopic scale [191]. A few basic geometries have been shown to achieve large responsivity and exceptional mechanical quality. Most widely, flexural modes of single clamped cantilever resonators or double clamped beam resonators are studied (see Fig. 3.1). Decoupling from modes in orthogonal dimensions and minimization of the oscillator mass suggest a nearly one dimensional geometry, characterized by a length l that is much larger than the transverse dimensions w, h. Alternatively, membrane oscillators with $l, w \gg h$ and bulk oscillators with $l \sim w \sim h$ can have amenities despite their larger mass. Tensile stress is an important parameter for doubly clamped beams or membranes, affecting the spectrum of resonance frequencies, the mode function, and also the mechanical dissipation. The length scale for micromechanical resonators continuously extends from mm scale down to molecular scale nano-oscillators like carbon nanotubes (CNT) or graphene ribbons.

3.1.1. Modefunction, resonance frequency, and effective mass

A solid object consisting of A atoms has $3A$ modes of oscillation. While the high frequency modes with wavelengths on the order of the interatomic distance are typically referred to as *phonons*, the modes with wavelengths on the scale of the dimensions of the device are termed *mechanical modes* [192]. In contrast to phonons, mechanical modes are strongly affected by the geometry of the device.

3.1 Fundamental properties

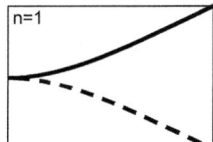
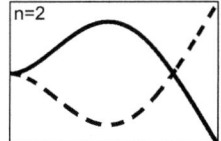
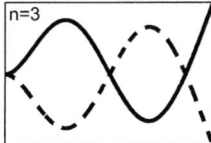

Figure 3.2.: Mode functions of the single clamped beam.

The mode shapes and frequencies can be accurately described by a continuum theory [190, 50] unless the device extensions shrink to molecular scales. Let us consider a beam resonator with length l much larger than the width w and thickness h, see Fig. 3.1 (a) and (b). The relevant material properties are the mass density ρ and the Youngs Modulus E. The dynamics of the transverse displacement along the z-axis, described by the mode function $u(x,t)$ extending along the beam axis x, obeys the differential equation

$$\rho w h \frac{\partial^2 u(x,t)}{\partial t^2} = -EI \frac{\partial^4 u(x,t)}{\partial x^4} \tag{3.1}$$

with $I = wh^3/12$ the bending moment of intertia of the beam. The general solution of this equation has the form

$$u_n(x,t) = [a_n \cos k_n x + b_n \cosh k_n x + c_n \sin k_n x + d_n \sinh k_n x] e^{-i\omega_n t}. \tag{3.2}$$

The clamping of the beam defines the boundary conditions and thereby the coefficients (a_n, b_n, c_n, d_n) and the eigenvalues k_n.

For a single clamped beam, the transverse displacement as well as the derivatives at the clamping point are zero, $u(0) = u'(0) = 0$, while the free end is force free such that $u''(l) = u'''(l) = 0$. Inserting these conditions into Eq. 3.2 yields a conditional equation for the eigenvalues k_n which can be solved numerically. Furthermore, the coefficients in Eq. 3.2 are found to be $a_n = -b_n$, $c_n = -d_n$, and numerical evaluation yields $c_n/a_n = -0.7341, -1.0185, -0.9920$ for the three lowest eigenmodes with $n = 1, 2, 3$. Figure 3.2 shows the first three modes of the single clamped beam.

The eigenvectors with the numerical values $k_n l = 1.875, 4.6941, 7.8548$ together with Eq. 3.1 determine the mode frequencies $\omega_n = \sqrt{EI/\rho w h} k_n^2$. The resonance frequency of the fundamental mode can be expressed by the material and geometry parameters, reading

$$\omega_1 = 2\pi \times 0.1615 \sqrt{\frac{E}{\rho} \frac{h}{l^2}}. \tag{3.3}$$

For comparison, a double clamped beam with otherwise identical geometry has a 6.4 times higher resonance frequency.

Effective mass

An important concept for the description of mechanical resonators is the introduction of the effective mass. When dealing with one peculiar mode of the oscillator, it is desirable to think of the resonator as behaving like a point mass concentrated at a position x_0 where e.g. the mode function has its maximal value. This facilitates the calculation of dynamic properties like the potential or kinetic energy. The mass of this point is the effective mass that contributes to the motion in the mode. It is calculated by [57]

$$M_{\text{eff}} = \frac{\rho w h \int_0^l u_n^2(x) dx}{\left(\int_0^l u(x) s^2(x - x_0) dx\right)^2} \quad (3.4)$$

where $s(x - x_0)$ denotes a probing function, describing e.g. the extension and profile of a readout or coupling laser. When probing is approximately pointlike (laser beam diameter $\ll l$), $s^2(x - x_0) = \delta(x - x_0)$, and for the fundamental mode Eq. 3.4 yields $M_{\text{eff}} = 33/140 \, M = 0.236 \, M$ for a single clamped beam, $M_{\text{eff}} = 0.735 \, M$ for a doubly clamped beam, and $M_{\text{eff}} = 1/2 \, M$ for a beam under high tensile stress.

3.1.2. Thermal motion

At finite temperature, all mechanical modes of oscillation of a mechanical resonator are thermally excited, including high frequency phonons. According to the equipartition theorem, each mode at frequency ω_m is carrying the Boltzmann energy $(1/2) M_{\text{eff}} \omega_m^2 a_{\text{th}}^2 = (1/2) k_B T$ in thermal equilibrium. When studying only a single mode, it is useful to describe the respective mode as a partially isolated degree of freedom and to treat all other modes of the oscillator and the support to which the resonator is clamped as a thermal bath. This thermal bath leads to a Langevin noise force F_{th} with a white spectral density [190, 50]

$$\langle S_{FF}(\omega, T) \rangle = \hbar M_{\text{eff}} \kappa \omega \coth\left(\frac{\hbar \omega}{2 k_B T}\right) \approx 2 k_B T M_{\text{eff}} \kappa, \quad (3.5)$$

where the approximation is valid for high temperatures $T \gg \hbar \omega / k_B$, and we have introduced the mechanical damping rate κ (see below for an explicit definition). This drives the oscillator into thermal motion with a mean number of phonons

$$n_{\text{th}} = \frac{1}{e^{\hbar \omega_m / k_B T} - 1}. \quad (3.6)$$

The resulting thermal amplitude can be obtained either directly from the equipartition theorem or by evaluating the Lorentzian response of the resonator to the thermal noise force,

$$a_{\text{th}}^2 = \frac{\langle S_{FF} \rangle}{M_{\text{eff}}^2} \int \frac{1}{(\omega_m^2 - \omega^2)^2 + \kappa^2 \omega^2} \frac{d\omega}{2\pi} \simeq \frac{k_B T}{M_{\text{eff}} \omega_m^2}, \quad (3.7)$$

3.1 Fundamental properties

where we have assumed that S_{FF} is sufficiently flat in the frequency range $\omega_m \pm \kappa$ of the resonance. This equation permits to evaluate the resonator response also for other noise sources acting on the resonator. The fluctuations in the amplitude are of the same size as the mean amplitude, and the oscillator changes its amplitude and phase on a timescale that is given by the inverse of the damping rate κ^{-1} of the mechanical mode. The state of the oscillator can be written as an incoherent mixture of coherent states $|\alpha\rangle$ with a gaussian probability distribution

$$p(|\alpha\rangle) = \frac{1}{\pi n_{\text{th}}} e^{-|\alpha|^2/n_{\text{th}}}. \tag{3.8}$$

With $|\alpha|^2 = n$ this is equivalent to an exponential phonon number distribution $p(n) \sim e^{-n/n_{\text{th}}}$.

For low temperatures $T \ll \hbar\omega_m/k_B$, the mean thermal occupation n_{th} becomes less than unity and the resonator resides mostly in its ground state. It then displays quantum fluctuations in its position with an rms amplitude

$$a_{\text{qm}} = \sqrt{\frac{\hbar}{2M_{\text{eff}}\omega_m}}. \tag{3.9}$$

E.g. for low mass and low frequency Si cantilevers (see Table 3.1, e.g. [62]) these fluctuations amount to about one picometer.

3.1.3. Dissipation

One corner stone for the large success of micro- and nanomechanical systems is the capability of achieving extremely low mechanical dissipation. Dissipation includes all processes that propagate energy stored in a single mechanical mode to other modes or excitations, be it the resonators thermal bath of other modes or some external environment like air. The figure of merit that quantifies dissipation is the quality factor Q, which states the number of oscillations the resonator undergoes until the energy stored in the oscillation has decayed to $1/e$ of the initial value. It is easily experimentally accessible, and can be infered either from the spectral width of a resonance of mechanical oscillations,

$$Q = \frac{\omega_m}{2\kappa}, \tag{3.10}$$

where ω_m is the resonance frequency and κ the FWHM of the Lorentzian amplitude spectrum, or from the amplitude decay time constant τ of an initially excited oscillation

$$Q = \frac{\omega_m \tau}{2}. \tag{3.11}$$

The highest mechanical quality factor obtained so far amounts to 2×10^9, measured for a cm scale single-crystal Si rod with $\omega/2\pi = 20$ kHz [193]. As a general trend,

Q is found to decrease with the size of the device, and micro- and nanomechanical resonators with Q substantially above a few thousands have been demonstrated only recently. Table 3.1 gives an overview of the parameters of state of the art micro- and nanoscale resonators with high quality factor.

Both fundamental and technical origins for mechanical dissipation have been identified. The individual sources of dissipation add up according to

$$Q^{-1} = Q_{\text{gas}}^{-1} + Q_{\text{ted}}^{-1} + Q_{\text{cl}}^{-1} + Q_{\text{def}}^{-1} + Q_{\text{sur}}^{-1}, \tag{3.12}$$

with the following contributions:

Gas damping: A resonator immersed in a gas such as air undergoes collisions with the gas atoms or molecules. This gives rise to damping [194] and to a noise force originating from adsorption and desorption processes [190]. For low pressure, gas damping is negligible, whereas one observes damping that grows linearly with the pressure $Q_{\text{gas}}^{-1} \propto p$ for pressure in the mbar range, characteristic for the free molecular flow regime. At even higher pressures, the resonator enters the viscous regime where $Q_{\text{gas}}^{-1} \propto \sqrt{p}$. The respective pressure for the transition between the three regimes depends non-trivially on the geometry and eigenfrequency of the mode under study.

Thermo-elastic damping: The restoring force of an acoustic mechanical oscillation is provided by a strain field. Due to nonlinear interaction with the surrounding bath of thermally excited phonons, energy will be transferred from the strain field to the bath [188]. In the diffusive regime, where the phonons thermalize fast compared to the timescale of the acoustic oscillation, the phonons constitute a temperature field, and the interaction between the mode and the bath can be described solely by the material's thermal expansion coefficient $\alpha_{th} = l^{-1} \partial l / \partial T$. Thermo-elastic damping thus arises from the coupling of the strain field to a temperature field, and from irreversible heat flow that is driven by the temperature gradients in this field. This damping process constitutes a fundamental limit, which can be expressed by a limiting value of the product of the quality factor and the mode frequency $Q_{\text{ted}} f_m$. High stress SiN strings [54] or membranes [55] approach this limit closely.

Clamping loss: At the clamping points of the oscillator to the support, vibrational energy is radiated away into the support. Calculations [195, 196] suggest that clamping loss is related to the relative size of the clamping contact, $Q_{\text{cl}} \propto l/w$, the number of clamping points, and the overall geometry. A related loss mechanism is energy transfer to other acoustic modes with comparable resonance frequency and large coupling to the considered oscillator mode [197].

Defects: Amorphous materials like fused silica or SiN contain a high density of defect states. Such defects can be successfully modeled by tunneling two-level systems formed by a double well potential [204]. In particular at low temperatures of a few Kelvin, where these systems become unsaturated, energy is coupled into these degrees of freedom and dissipation increases significantly [205, 189]. The stoichiometry of the material has proven to be crucial e.g. for the dissipation in SiN films

3.2 Detection and manipulation of mechanical motion

type	dimension (h, w, l) [µm]	Q-factor	$\omega_m/2\pi$ [Hz]	mass [kg]	T [K]	
Si cantilever	(0.17, 5,50)	2.5×10^5	9×10^4	10^{-13}	300	[198]
Si cantilever	(0.1, 3,120)	4.4×10^4	4×10^3	10^{-13}	4.2	[199]
SiN membrane	(0.05,500,500)	4×10^6	1×10^7	10^{-11}	300	[55]
SiN membrane	(0.05,1000,1000)	1.2×10^7	1×10^5	10^{-10}	2	[200]
SiN strings	(0.11, 0.35, 300)	1.2×10^6	1×10^6	10^{-14}	300	[54]
SiC nanowire	(0.2, 0.2,130)	1.6×10^5	3×10^4	10^{-14}	300	[201]
SWCNT	(0.002, 0.002, 0.3)	1.4×10^5	3×10^8	10^{-21}	2	[202]
Graphene sheet	(0.001, 0.2, 2)	1.4×10^4	5×10^6	10^{-22}	5	[203]

Table 3.1.: Examples for state of the art micro- and nanomechanical resonators with high quality factors. The given temperatures refer to the conditions for the determination of the quality factor. Single wall carbon nanotubes (SWCNT) are further discussed in chapter 6.1.

[55, 206]. It is conjectured that high tensile stress considerably changes the energy distribution of defect states and thereby restores high mechanical Q at cryogenic temperatures [206] for amorphous materials.

Surface quality: Oxide layers, metallization layers, adsorbed liquids or solids, and surface defect states like dangling bonds are known to increase mechanical damping. By annealing of ultra-thin Si resonators [52, 198], mechanical loss could be reduced by up to two orders of magnitude. As a general trend, the quality factor is observed to decrease for smaller structures with larger surface to volume ratio.

3.2. Detection and manipulation of mechanical motion

Optical detection and manipulation of mechanical motion has developed rapidly in the recent years and established the new field of cavity optomechanics [67, 65, 66, 68]. The sensitivity and control has achieved a level where position fluctuations can be monitored (almost) down to the standard quantum limit [56, 72, 207, 69, 71, 70]. The standard quantum limit refers to the ultimate sensitivity achievable in a weak, linear, and continuous measurement process [208]. It arises from the well known fact that a measurement always disturbes the measured object, often called measurement back action. Thus one has to find a compromise between measuring weaker but gaining less information about the object and measuring stronger but imparting excessive back action noise on the oscillator. The optimal setting is found where measurement imprecision noise S_{zz} (dominated by photon shot noise at the detector) and back

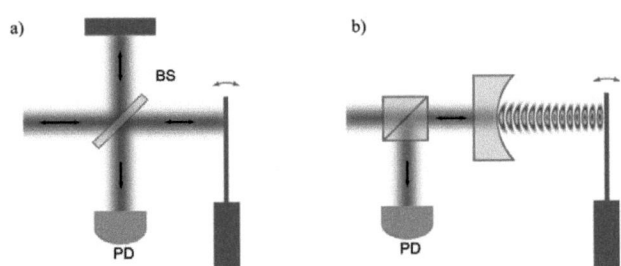

Figure 3.3.: Optical readout schemes. (a) Michelson-Interferometer with a mechanical resonator in one arm. (BS) 50:50 beamsplitter, (PD) photodetector. (b) Optical cavity incorporating a mechanical resonator as one end mirror.

action noise $S_{\text{FF,ba}}$ (dominated by radiation pressure noise acting on the oscillator) are equal and add up to a total inferred position noise $\langle S_{\text{tot}} \rangle = \langle S_{zz} \rangle + \langle S_{\text{FF,ba}} \rangle = 2\langle S_{\text{qm}} \rangle$ which is twice the noise power of the zero-point fluctuations S_{qm} of the mode under study. The minimum achievable position uncertainty is thus

$$\Delta z_{\text{QL}} = \sqrt{2} a_{\text{qm}} = \sqrt{\frac{\hbar}{M_{\text{eff}} \omega_m}}. \tag{3.13}$$

Detection of such extremely small amplitudes is achieved by interferometric techniques. A standard tool is the Michelson interferometer, where a mechanical resonator is incorporated as an end mirror in one arm. The differential length change of the two arms leads to measurable intensity modulation at the output port (cf. Fig. 3.3 (a)). Sensitivities of $\sim 10^{-15}$m/$\sqrt{\text{Hz}}$ are possible with this technique [209, 210], and with modifications like introducing enhancement cavities inside the interferometer arms as used e.g. for gravitational wave detection, sensitivities of 10^{-19}m/$\sqrt{\text{Hz}}$ have been achieved [72, 211].

The devices for ultimate optical position readout are optical cavities that integrate a mechanical oscillator either as one of the cavity mirrors [72, 71] (cf. Fig. 3.3 (b)), as an additional device inside the cavity [200, 76], or as integral degree of freedom of a bulk optical microcavity [69, 70]. Oscillator displacement affects the resonance condition of the cavity and thus establishes a direct coupling between the mechanical motion and the light field inside the cavity. On the one hand, the resulting phase or amplitude modulation can be used to achieve exceptional readout sensitivity down to 10^{-19}m/$\sqrt{\text{Hz}}$ [56, 72]. On the other hand, radiation pressure of the light reflected of the resonator exerts a force on the oscillator and thus provides a back action that can be harnessed to cool [200, 72, 207, 69, 71, 70] or parametrically amplify [212] mechanical motion. Employing such back action cooling, cryogenically precooled oscillators have been cooled down to a few tens of phonons remaining in the respective mode [69, 71, 70, 207], and e.g. a record value $T_{\text{eff}} = 1.4$ μK for the effective temperature of the pendulum mode of a 2.7 kg mirror has been achieved [72].

However, the oscillators used in these experiments are rather large since they are designed to support an entire optical mode. Experiments studying truely nanomechanical motion have been mainly performed in a solid state setting, e.g. by capacitively coupling the resonator to a single electron transistor [213, 214, 215] or a microwave cavity [216, 207, 58]. There, a sensitivity close to or even below the standard quantum limit has been already achieved [207, 58]. Recently, novel approaches also achieve optical nanomechanical motion detection with excellent sensitivity, either by coupling a nanoresonator to the evanescent field of a microtoroid [57], by introducing a nanoresonator into an optical cavity [76], or by the design of optically resonant structures integrated into the mechanical oscillator [217].

3.3. Quantum states of mechanical oscillators

The creation and observation of quantum effects in a single mode of a mechanical resonator is a central goal that is shared by the communities of cavity optomechanics [66, 67], micro- and nanoelectromechanical systems (MEMS/NEMS) [192], and gravitational wave detection [218, 219]. The motivation for this is manifold. Firstly, it is the hallmark of ultimate control over a system. Practically, it would have implications for the achievable sensitivity e.g. in force sensing [51], gravitational wave sensing [218], or nuclear spin imaging [63]. Secondly, it would provide a new testbed for non-standard decoherence theories [220, 221] and illucidate the crossover from microscopic quantum systems to macroscopic classical systems. Finally, mechanical resonators coherently coupled to other controlled quantum systems could establish a new class of coupled systems at the quantum level, so-called *hybrid quantum systems*. This is interesting in the context of quantum engineering, quantum information networks, and quantum information processing.

The systems investigated in the different fields have individual advantages and drawbacks. Optomechanical systems are so far realized by a linear coupling of mechanical motion to a strongly driven and thus classical optical cavity mode. In this limit, the coupling can be very strong, and e.g. a mechanical analog of the Autler-Townes splitting in the energy spectrum of a coupled resonator-cavity system was demonstrated [222]. Theoretical proposals have shown that the coupled system could generate mechanical squeezing [223, 224], entanglement between the light field and the resonator [225, 226], or the creation of Schrödinger cat states [227, 226]. However, the creation of mechanical Fock states or more general quantum superpositions are not accessible by a linear coupling to a classical field, and coupling to e.g. a two-level system would be required. Alternatively, dispersive measurement concepts that rely on a quadratic coupling to the light field could permit the observation of quantum jumps [200, 228]. Finally, non-classical light states such as superpositions of zero- and single-photon states could be employed to generate superpositions of position states [229, 230]. Yet, single photons lead only to a weak coupling, and the required experimental conditions for such experiments are hard to reach.

Mechanical resonators in an entirely solid-state based environment certainly have the longest tradition, and quantum manipulation of resonators has been discussed for a variety of systems [231, 232, 233, 62, 234, 235]. NEMS systems have already been successfully coupled to two-level systems such as a single electron spin in a solid [62] or a cooper pair box [236]. In the latter experiment, the quantum state of the two-level system, i.e. the presence or absence of an extra Cooper-pair in the box, was read out by a nanomechanical resonator [236]. Very recently, an experiment of similar kind has demonstrated quantum control of a mechanical oscillation [237] coupled to a qubit. There, a 6 GHz oscillation of the thickness of a piezoelectric slab has been cryogenically cooled to the ground state. By strongly coupling this mode to a quantum controlled Josephson phase qubit, single phonon states could be prepared in the mechanical resonator. Furthermore, the observation of Rabi oscillations between the resonator and the qubit for a single quantum demonstrated the achievement of the strong coupling regime. This experiment displays the big advantages of entirely solid-state based approaches, being in essence a natural compatibility with cryo-technology and high achievable coupling rates (e.g. $2g_0 = 123$ MHz in the experiment above,[237]). Yet, the coherence time of solid state based two-level systems as used so far reside in the ns to ms regime (e.g. 13 ns in [237]) and thus require exceptional high mechanical resonance frequency. Furthermore, the mechanical resonator is typically integrated in a complex electronic environment and carries electrodes, making it difficult to independently optimize the mechanical quality factor.

Using ultracold atoms to couple to mechanical oscillators would create a novel, qualitatively different setting. Atomic systems would contribute the advantage of exceptionally long coherence times. Superpositions of internal states have been demonstrated to preserve coherence for several seconds [10, 11] also in close proximity to surfaces. In addition, an elaborate toolbox for the quantum manipulation of the atomic state is at hand [1, 8, 9] giving control over internal and external degrees of freedom. Atoms could provide both the continuous degree of freedom of collective atomic motion, and a discrete set of internal levels that can be reduced to a two-level system. Furthermore, a unique feature of ultracold atoms is that dissipation can be tuned, e.g. by switching on/off laser cooling, and that self-interactions can be tuned, e.g. with Feshbach resonances [238, 239, 240]. In a large set of theoretical proposals [35, 36, 37, 38, 39, 40, 41, 42, 43, 44, 45, 46, 47, 48], various schemes for the coupling of mechanical motion to different atomic degrees of freedom have been discussed. This illustrates that the hybrid approach is multifaceted and permits to individually optimize the separate systems and the interaction mechanism. Yet, the experimental realization is involving and requires to combine the control of ultracold atoms with an cryogenic environment. In the outlook 6 we discuss three coupling

3.3 Quantum states of mechanical oscillators

scenarios that have the potential to achieve atom-oscillator coupling at the quantum level.

3.3.1. Decoherence

Quantum states of a mechanical oscillator will show decoherence due to interaction with the environment at a rate that is in general faster than the damping rate κ of the amplitude of the oscillator. The standard derivation of the decoherence rate assumes that the environment can be modelled as an ohmic bath of harmonic oscillators [241]. The interaction is described by a linear coupling in the position of the mechanical oscillator z_c with coupling strengths g_i

$$H_{\text{int}} = z_c \sum_i g_i q_i, \tag{3.14}$$

where q_i are the coordinates of the bath harmonic oscillators. In the quasi classical or high temperature limit for the bath, where only thermal excitations of the field are taken into account (and zero point fluctuations are neglected), one can derive a master equation for the density matrix of the oscillator. In position space it reads [241, 242, 243]

$$\begin{aligned}\frac{d\rho(z_c, z'_c)}{dt} = &-\frac{i}{\hbar}[H_0, \rho(z_c, z'_c)] - \kappa(z_c - z'_c)\left(\partial_{z_c} - \partial_{z'_c}\right)\rho(z_c, z'_c) \\ &- \frac{2m\kappa k_B T}{\hbar^2}(z_c - z'_c)^2 \rho(z_c, z'_c).\end{aligned} \tag{3.15}$$

The first two terms describe the coherent dynamics according to the system Hamiltonian of the mechanical oscillator $H_0 = p_m^2/2M_{\text{eff}} + (1/2)M_{\text{eff}}\omega_m^2 z_c^2$ and the damping of the oscillator amplitude respectively. These are the relevant terms for the (near) diagonal elements of the density matrix where $z_c \approx z'_c$. The third term describes thermal fluctuations coupling to the oscillator. It acts mostly on the off-diagonal elements of ρ and thus reduces the coherences.

For an oscillator in a superposition of two position states with distance Δz_c, the off-diagonal elements will be peaked around two maxima at a distance $(z_c - z'_c) \approx \Delta z_c$, so that the prefactor of the third term, which is found to be the decay rate of an exponential damping of the coherences, can be written as [241, 242, 243]

$$\gamma_{\text{dec}} = \frac{2m\kappa k_B T}{\hbar^2}(\Delta z_c)^2 = \frac{k_B T}{\hbar Q}\left(\frac{\Delta z_c}{z_{\text{QL}}}\right)^2 = \kappa n_{\text{th}}\left(\frac{\Delta z_c}{a_{\text{qm}}}\right)^2. \tag{3.16}$$

We have used $\kappa = \omega/2Q$ and the quantum limit of continuous position measurement $z_{\text{QL}} = \sqrt{2}a_{\text{qm}}$. This result is commonly referred to as the *golden rule of decoherence*. Its validity is restricted to small decoherence rates, either due to small separation

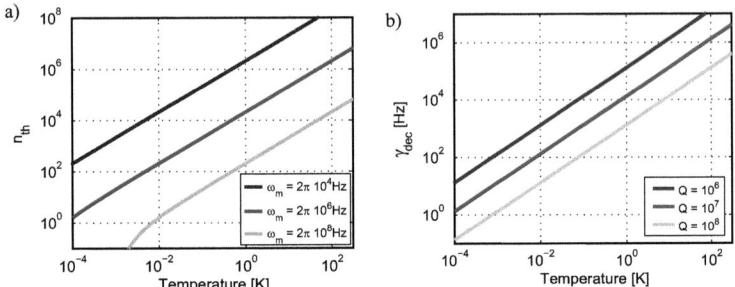

Figure 3.4.: (a) Thermal phonon occupation of the resonator mode. Cyrogenic cooling of high frequency oscillators can achieve $n_{\text{th}} \sim 1$. (b) Decoherence rate γ_{dec} for a Fock state or a superposition with $\Delta z_c = a_{\text{qm}}$. A state of the art resonator with $Q = 10^7$ [200] put into the coldest present day cryostat with $T < 10^{-4}$ K [244, 245] would show a coherence lifetime of one second.

Δz_c or small damping κ, such that $\gamma_{\text{dec}} \ll \omega$ and the oscillator can undergo many cycles before it decoheres. For larger amplitudes, the dynamics of the bath becomes important and the decoherence dynamics is no longer exponential [246].

In Figure 3.4 (a) we show the thermal phonon number according to Eq. 3.6 for resonance frequencies spanning current micro- and nanoresonators with demonstrated high quality factor. The temperature represents the range presently achievable by cryogenic cooling techniques [244, 245]. For resonance frequencies $\omega_m/2\pi \gtrsim 1$ MHz, cryogenic ground state cooling is in reach. Figure 3.4 (b) shows the decoherence rate given in Eq. 3.16 for a single phonon Fock state or a superposition $|\Psi\rangle = 2^{-1/2}(|0\rangle + |\alpha = 1\rangle)$. Remarkably, present day mechanical quality factors in combination with state of the art cryogenic cooling should achieve coherence lifetimes on the order of one second.

Finally, we mention that non-standard decoherence models such as the objective reduction model [247, 221] that predicts a decoherence rate that is proportional to the gravitational self energy, $\gamma_{\text{dec}} \sim \Delta E_{grav}/\hbar$ could be tested with mechanical resonators. A mechanical oscillator with $Q \sim 10^9$ at $T \sim 10^{-5}$ K [230] would enter a regime where decoherence due to gravitational interaction would be faster than the rate predicted by Eq. 3.16.

3.3.2. The size of a superposition

Since one of the motivations to push mechanical objects towards quantum behaviour is based on the question, how big quantum states can be and if there are any unconventional mechanisms for decoherence, we sketch here some considerations about

3.3 Quantum states of mechanical oscillators

the size of a superposition.

The size of a superposition is not obviously defined. Dür et al [248] proposed a measure of the size that includes the effect of state overlap. Consider a superposition of N particles that can be described by

$$|\psi\rangle = \frac{1}{\sqrt{K}} \left(|\phi_1\rangle^{\otimes N} + |\phi_2\rangle^{\otimes N} \right) \quad (3.17)$$

where $K = 2 + \langle\phi_1|\phi_2\rangle^N + \langle\phi_2|\phi_1\rangle^N$ and $|\psi\rangle$ is a state of N two-level systems. In general, $|\phi_1\rangle$ and $|\phi_2\rangle$ are not orthogonal but have finite overlap $\langle\phi_1|\phi_2\rangle^2 = 1 - \zeta^2$, e.g. $\zeta^2 = 1 - \exp(-|\alpha - \alpha'|^2)$ for a superposition of coherent states $|\alpha\rangle, |\alpha'\rangle$. From an analysis of the decoherence rate or alternatively the entanglement content of such a state, the authors of Ref. [248] derive the effective size

$$\mathcal{S} \sim \zeta^2 N, \quad (3.18)$$

which allows a direct comparison with a pure GHZ state $|\psi\rangle = 2^{-1/2}(|0\rangle^{\otimes N} + |1\rangle^{\otimes N})$, often called a *Schrödinger cat state*. Extensions to general manybody states were derived recently, either by considering the average number of single-particle operations necessary to map one part of the superposition to the other [249] or by asking how many fundamental subsystems of the object have to be measured to project the state into a single branch [250]. For states of the form 3.17, Eq. 3.18 is reproduced.

However, these measures do not consider the number of excitations making up the superposition and also treat compound objects irrespective of their mass. A measure that is closer to our intuitive understanding of macroscopicness could be related to the energy difference of the individual parts of the superposition, e.g. for a mechanical resonator in a superposition of two position states $\Delta E \sim M_\text{eff} \omega^2 \Delta z_c^2$. With normalization to obtain dimensionless values, it could read

$$\mathcal{S} = \zeta^2 \frac{M_\text{eff}}{m_p} \left(\frac{\omega_m}{\omega_0} \right)^2 \left(\frac{\Delta z_c}{a_\text{qm}} \right)^2, \quad (3.19)$$

where the eigenfrequency ω_m is normalized e.g. to $\omega_0 = 1$ Hz, and the effective mass according to Eq. 3.4 is normalized e.g. to the proton mass m_p. For collective states such as the dipole mode of an atom cloud, $M_\text{eff} = N m_\text{at}$. Using a_qm to normalize the displacement Δz_c gives a factor that also enters the decoherence rate in Eq. 3.16 and corresponds to the number of phonons involved in the superposition. In line with the argumentation in [248], the modified size estimate Eq. 3.19 is proportional to the decoherence rate of the state.

If one compares this measure for a single atom in a superposition between being at rest and oscillating with an amplitude $\Delta x = 10\ \mu$m in a trap with $\omega/2\pi = 100$ Hz as it was similarly realized in [9], and a carbon nanotube (CNT) with $M_\text{eff} = 2 \times 10^{-20}$kg $\hat{=} 10^6$ atoms and $\omega_m/2\pi = 20$ kHz (see chapter 6.1) in a superposition of two position states which differ by the CNT ground state spread $\Delta z_c = a_{\mathrm{qm}} = 0.2$ nm, one obtains $\mathcal{S}_{\mathrm{CNT}}/\mathcal{S}_{\mathrm{at}} \sim 10^7$. This is in contrast to the size estimate given in Eq. 3.18, which yields $\mathcal{S}_{\mathrm{CNT}}/\mathcal{S}_{\mathrm{at}} \sim 1$ for this case.

4. Setup and BEC production

In this chapter I describe and characterize the experimental setup and summarize the techniques used for the production of Bose-Einstein condensates.

The system was designed along the lines of previous experiments as described in [82, 150] and has been built up from scratch. The core of the experiment is an atomchip with an integrated microcantilever. The chip is an integral part of a glass cell which is attached to a compact, ultrahigh vacuum system. A diode laser system is used for laser cooling and detection of atoms. The vacuum cell is surrounded by three pairs of Helmholz coils to provide homogeneous magnetic bias fields. The coils and the chip wires are supplied by low-noise, fast switchable current sources, which provides versatile magnetic fields for the trapping of atoms.

For the experiments described in chapter 5, where we couple the collective motion of a cloud of atoms to the oscillations of a microcantilever via short-range surface forces, high control over the motional state of the atoms and small spatial extent of the clouds is crucial. In a BEC, the atoms are prepared in the motional ground state and thus show minimal position and momentum spread. With our setup we achieve fast, robust, and reproducible preparation of BECs and precise positioning of the atoms close to the microcantilever.

Using surface forces to couple BECs to the cantilever motion requires close approach of the atoms to the cantilever. In the last section, we characterize the BEC lifetime and investigate limiting effects for the atoms in magnetic traps close to the cantilever surface.

4.1. Atom chip with microcantilever

In collaboration with D. Anderson, our group has developed a very compact solution for the integration of an atom chip to a vacuum system [251]. The chip is used to seal the vacuum by gluing it on an open side of a glass cell. This minimizes the size of the glass cell, the chip itself serves as an electrical feedthrough, and optimal optical access is provided.

The presence of the chip calls for a special solution for the geometry for laser cooling. The chip blocks the optical access for one half-space and the usual six-beam configuration for the operation of a Magneto-Optical-Trap (MOT) would require quite large separation from the chip. Instead, we use the mirror MOT configuration [252] to operate a MOT close to the chip. It replaces two cooling beams by the

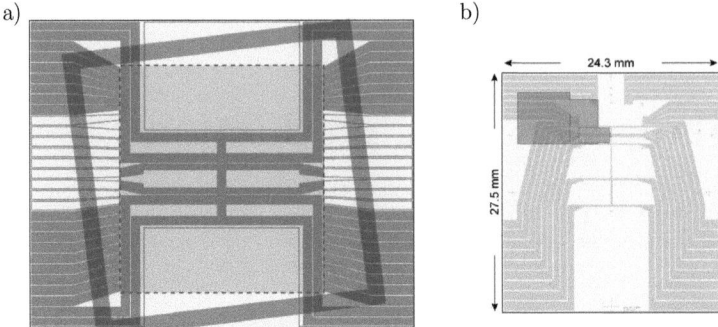

Figure 4.1.: Chip wire layout. (a) Base chip with wires for intermediate MOT phases and feedthrough wires for the microtrap chip. The outline of the rotated glass cell and the position of the experiment chip (dashed line) are shown. (b) Microtrap chip with wires for magnetic micro traps. The outline of the cantilever subassembly is indicated.

reflection of two beams inclined by 45° on a mirror that is applied on the chip. In this configuration, laser cooled atoms can be directly trapped with the magnetic microtraps created by the chip, without the need of additional trapping and transport stages.

The integration of a mechanical oscillator on an atomchip calls for a seperate cooling region suitable to run a mirror MOT, and an experiment region where the cantilever is mounted.

4.1.1. The atom chip

It has proven to be advantageous to use a stack of a so-called *base chip* with macroscopic (mm-scale) wires for the use during MOT phases, and a second, so-called *microtrap chip* with wires for magnetic trapping. The base chip seals the glass cell and provides electrical contact to the outside and to the microtrap chip. This design is advantageous for absorption imaging, as the microtrap chip inside the vacuum introduces a spacing between the imaging axis and the glue meniscus that forms when the base chip is glued to the glass cell. Figure 4.1 shows the wire layout of the two chips.

Base chip

The base chip has a size of 45×38 mm^2 and is fabricated on an 800 μm thick AlN ceramic substrate. This material is electric insulating, has a large thermal

4.1 Atom chip with microcantilever

conductivity of 180 W/Km, and good mechanical stability. The chip size is chosen to be large enough to attach a 35×35 mm^2 glass cell rotated by an angle of $7°$ on the chip. This avoids fringes and standing waves in the imaging beam. The contact pads at the sides of the chip are designed to match the pitch of a PCI connector, such that the wires can be directly contacted with a PCI socket connector. The central wires (not filled in fig. 4.1 (a)) remain unused and the PCI connector is carved out at these positions to provide optical access for imaging in the central region close to the chip. The outermost wires form "H" shapes (to allow for both "U" and "Z"-shapes in two possible orientations) in the center of the chip with a minimum wire width of 1 mm. Currents in this wires contacted for U-shape provide quadrupole fields during a phase where the MOT is ramped close to the chip.

For chip fabrication we use an Au electroplating process (see Appendix D). In brief, we evaporate a 3 nm thick Ti adhesion layer and a 50 nm thick Au layer on the entire chip surface and define the wire geometry by photolithography with a foil mask. The metallization serves as a seed layer for the subsequent electroplating process that grows the wires to a thickness of 10 μm. The wires on this chip can carry maximal currents of up to 10 A.

Microtrap chip

The microtrap chip has a size of 24.3×27.5 mm^2 and is fabricated on the same type of substrate. The size is chosen to fit inside the glass cell including wire bonds, and to allow for the rotation of the cell. A wire width comparable to the smallest atom-wire distance is sufficient to avoid field reduction by the finite size of the wire. As our experiments are performed close to a surface but far from the wires on the chip, the wires can be relatively large. The connection pads that serve for wire bonding have a width of 1 mm, and the wires are tapered down in the center of the chip to a smallest width of 50 μm. We use the same fabrication process as described above, but use a chrome mask to achieve better resolution, and we grow the wires to a thickness of 5 μm.

We test the power stability of the chip wires by sending a constant current through the wire and recording the voltage drop accross the wire, thus measuring the resistance. A change in resistivity indicates resistive heating and allows one to estimate the maximum current the wire can stand. When sending a current of 3 A through our smallest wire with cross section 5×50 μm^2, we observe a relative change in the resistance of 5.7 %. This is still a small value, for typical relative changes for wire breakdown are of the order of 50 % [253]. From a simple estimate for the contribution of the resistance at the thin part of the wire, and using the temperature coefficient $\alpha_T = 4.0 \times 10^{-3}$ K^{-1} for gold, we calculate a temperature increase of the wire of 21°C. The current density in this case amounts to $j = 1.2 \times 10^{10}$ A/m^2, sufficiently below the typical breakdown current density for thin wires of $j_{\max} = 10^{11}$ A/m^2 [253].

Setup and BEC production

The wire layout comprises three main elements. First, a 3 mm wide H-geometry with 200 μm wires for large currents. It is centered to the MOT quadrupole wires on the base chip and is used to create a large volume trap for initial trapping of laser-cooled atoms to load the atomchip. Second, a wire guide tapering down from 200 μm to 50 μm width is used to transport the atoms between the loading region and the position of the cantilever. The distance to the cantilever has to be large enough to provide sufficient area for the MOT beams, and we choose an overall transport distance of 6.4 mm between the center of the first trap and the cantilever. At the location of the cantilever, a 2 mm wide H and four dimple wires of 50 μm width and with a pitch of 90 μm provide traps at several positions with strong confinement along all axes. Additional wires are used to contact a piezoelectric transducer.

Dielectric mirror

We apply a dielectric mirror on the chip for the operation of a mirror MOT. We use a transfer technique, where a detachable dielectric mirror coating supplied on a carrier substrate[1] is glued on the desired chip area. The coating has > 99% reflectivity at $\lambda = 780$ nm for both s and p-polarization under 45° incidence, but the different penetration depth of the two components will alter circular polarization slightly.

As the cantilever should be at a small, well defined distance above the wires, it is undesired to have a mirror there. For this reason we apply the mirror only on a part of the microtrap chip. The transfer substrate is cleaved to the desired size after scoring the coated side. The brittle coating is likely to sliver, and a corrugated stripe of typically $\sim 500\mu$m width results. We use Epo-Tek 353 ND after thorough outgassing to glue the mirror on the chip. Careful dosing (with several tests) is necessary, as excessive glue will form a large meniscus at the edge of the coating that may obstruct the transport of the atoms accross the chip. On the other hand, too little glue may result in air enclosures that act as virtual leaks, or the detachment of the coating from the glue in the space between wires during curing. In our case, a meniscus of < 100 μm height was achieved and proved to be sufficient not to influence the atomic transport. The volume reduction of the glue during curing and differences in the thermal expansion of the coating and the chip lead to fractures and deformations of the coating. However they do not affect the performance of the MOT.

Bonding of the chips

The microtrap chip is glued to the base chip by a UHV compatible, heat conductive Epoxy glue, typically Epo-Tek H77S. Complete filling of the gap between the chips is important to avoid virtual leaks, and a minimal thickness of the glue layer is desired for optimal thermal conductivity.

---

[1] We obtain it from the company O.I.B., 07745 Jena, Germany.

4.1 Atom chip with microcantilever

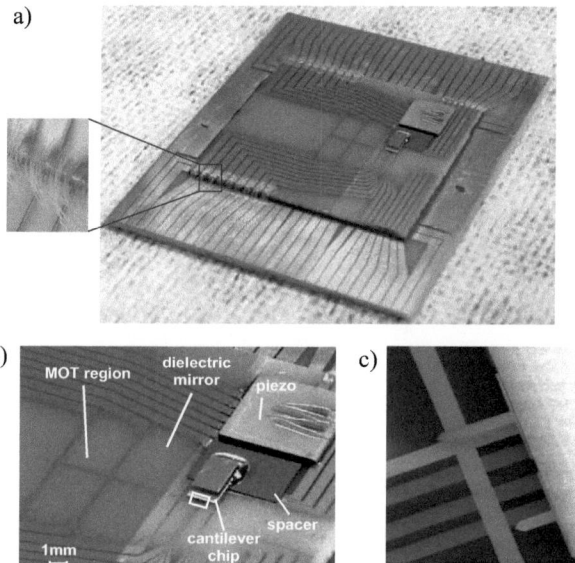

Figure 4.2.: (a) Photograph of the assembled chip. The experiment chip is glued on the base chip and contacted with wire bonds (see inset), the cantilever subassembly is glued on the experiment chip. (b) Close up of the experiment chip. The cantilever subassembly comprises a spacer chip, a piezo for actuation, and a cantilever chip with two cantilevers for experiments. (c) SEM micrograph of the two cantilevers above a wire cross used for trapping. The larger cantilever is used for the experiments reported here, the second smaller cantilever was intended for optical lattice experiments.

The electric contact between the two chips is provided by wire bonds. We use a Ball bonder and 25 μm gold bond wires that can carry up to 800 mA each. We apply 12 bond wires for chip wires that carry maximal currents of 3 A, and 8 bonds for low current wires to guarantee sufficient safety overhead.

Integration of the resonator

We integrate the mechanical resonator directly on the microtrap chip to achieve high control of the magnetic traps close to the cantilever and to minimize the mechanical complexity in the vacuum. To obtain a defined distance between cantilever and wires, we glue a 5×7 mm^2 large, 45 ± 5 μm thick SOI spacer chip on the atom chip. We integrate a 5 × 5 mm² large, 300 μm thick piezo-electric transducer on the spacer for mechanical excitation of the cantilever. Since the piezo sits on top of the spacer, excitation relies on inertially generated sonic waves in the chip rather than on the direct elongation of the piezo. We use Epo-Tek H20E to electrically contact the bottom side of the piezo to three wires, and bond wires to contact the top side to a neighbouring wire.

The cantilever is supplied on a Pyrex glass chip with a total size of 1.7 × 3.4 × 0.5 mm³ which contains three more cantilevers placed on two sides of the chip. For the final step of the gluing procedure, the cantilever chip is attached to a micropositioning mount to achieve precise alignment. Observed with a long working distance stereo microscope, we can align one of the cantilevers above a wire cross. For gluing, the cantilever chip is slightly pressed on the spacer chip, and the glue is applied at the backside of the chip to let capillary forces fill the space between the two chips. This avoids flow of the glue to the cantilever during the low viscous phase of the glue in the beginning of the curing. Defined by the spacer chip and the glue layers, the cantilever surface is located at a distance of 68 μm from the wire surface, with an uncertainty of ±10 μm due to the unknown thickness of the glue layers. From measurements with atoms (see chapter 5), this distance can be inferred with a precision of ±2 μm.

Figure 4.2 shows a photograph and a SEM image of the completed chip.

4.1.2. Resonator characterization and readout

The mechanical resonator is a commercial SiN cantilever (PNP-DB-2) which is typically used for AFM measurements[2]. It has specified dimensions of $(200, 40, 0.6)$ μm and carries a 65 nm thick Au/Cr coating on one side for optical readout. The specified fundamental resonance frequency is 16 kHz and its force constant is 0.05 N/m. We calculate the mass from the geometry and obtain $M = 20$ ng, a factor two larger than expected from the specified force constant, but in agreement with measurements of the cantilever eigenfrequency. From SEM images of a similar cantilever from the same batch we measure dimensions of $(195, 35, 0.43)$ μm.

Optical readout

We implement a standard beam deflection readout [254] that allows us to monitor the cantilever motion with simple optics from outside the vacuum cell. It is based on the angular deflection of a laser beam reflected on the tip of the cantilever. To permit readout also during atom preparation, we use a grating stabilized diode laser running free at $\lambda = 830$ nm, far detuned from atomic resonances. After mode cleaning in a singlemode fiber, the beam is collimated by an achromatic lens with

---

[2]We obtained it as a free test sample from NanoAndMore GmbH, 35578 Wetzlar, Germany

4.1 Atom chip with microcantilever

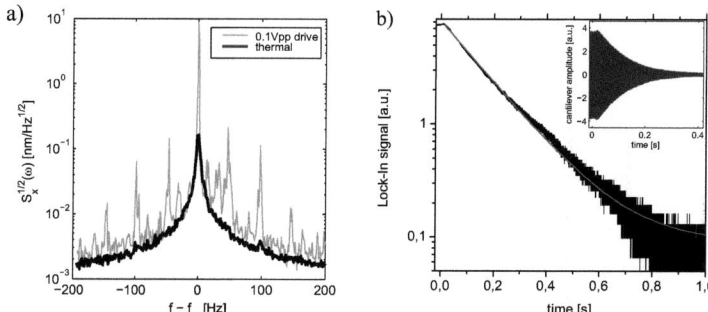

Figure 4.3.: (a) Amplitude spectrum $\sqrt{S_z(\omega)}$ of cantilever oscillations for thermal motion and resonant piezo excitation with 100 mVpp driving amplitude. The cantilever frequency here is $f_m = 9724$ Hz. The sensitivity of the readout is 2×10^{-12} m/$\sqrt{\text{Hz}}$. We calibrate the piezo excited oscillation amplitude by comparison with the thermal motion and find a driving efficiency of 80 ± 12 nm/Vpp. (b) Cantilever ringdown after an initial excitation to $a = 160$ nm measured with the LockIn-amplifier. The inset shows the oscillations observed with an oszilloscope. The measured time constant is $\tau = 0.11$ s, corresponding to a Q-factor of $Q = 3200$.

focal length $f = 80$ mm and focused with a single achromatic lens with $f = 160$ mm to a measured beam diameter of 35 μm. Two mirrors direct the beam on the tip of the cantilever, and the reflected beam is measured with a quadrant photodiode[3] that detects oscillations of the beam position with a bandwidth of 1 MHz.

The photodiode signal is observed either with an oscilloscope after AC-amplification, or the signal amplitude at the cantilever eigenfrequency is measured with a Lock-In amplifier[4] or with a spectrum analyzer[5].

To find the cantilever fundamental resonance we excite vibrations in the common support by the piezo on the spacer chip and scan the excitation frequency to obtain a spectrum of the amplitude response of the cantilever. The measured resonance frequency drifts from day to day by up to ~ 20 Hz, and we find a decrease from $\omega_m/2\pi \equiv f_m = 10350$ Hz to $f_m = 9530$ Hz within one year. The drift does not depend on whether experiments are performed, and we attribute it to aging of the layered Au/Cr/SiN structure. The quite large deviation from the specified frequency can be explained by the deviation from the specified thickness of the cantilever, and we reproduce the measured frequency in a FEM simulation of the cantilever for the dimensions obtained from SEM images. The simulation predicts higher order modes

---
[3]Thorlabs PDQ80A
[4]Stanford Research Systems SR830-DSP dual channel digital Lock-In amplifier, 1mHz - 102.4kHz
[5]Agilent E4405B 9 kHz - 13.2 GHz

at frequencies $(63.2, 179, 352, 586)$ kHz.

From measurements of the cantilever response at different excitation frequencies we determine the FWHM of the amplitude spectrum of the resonances to be $\kappa = 6.2$ Hz. However, the driving can influence the line shape, and to determine the mechanical quality factor we use a different method. We excite the cantilever to large amplitude and observe the damping of the amplitude after switch-off of the piezo drive. The exponential decay yields a time constant of $\tau = 0.11 \pm 0.02$ s, corresponding to a quality factor $Q = \omega_m \tau / 2 = 3200 \pm 600$. Figure 4.3 (b) shows a ringdown measurement taken with the Lock-In amplifier. The inset shows the time trace observed on an oscilloscope, where the individual oscillations can be seen (not resolved here).

Calibration of the cantilever amplitude

We can resolve the thermal motion of the cantilever with the Lock-In amplifier and thus observe the temporal fluctuations of the oscillations. Alternatively we can take a spectrum with the spectrum analyzer and observe the full Lorentzian shape of the fundamental resonance. We average over $10 - 30$ measurements to obtain the mean amplitude spectrum of the oscillations. For longer averaging times, frequency drifts spoil the measurement.

Figure 4.3 (a) shows a typical spectrum of the thermal motion, along with a spectrum taken for resonant piezo excitation with a driving amplitude of 100 mVpp. The additional weaker resonances at $\pm(47, 98, 145)$ Hz which become visible in the driven case probably come from the mains supply of the frequency generator.

We calibrate the amplitude spectrum by integrating over the power spectral density of the thermal motion and setting it equal to the square of the calculated thermal motion amplitude,

$$\int S_z(\omega) d\omega = a_{th}^2 = \frac{k_B T}{M_{\text{eff}} \omega_m^2}, \qquad (4.1)$$

with the effective mass $M_{\text{eff}} = 0.24\,M$.

With this calibration we can infer the amplitude for piezo excited oscillations. We find a driving efficiency of 80 ± 12 nm/Vpp. The error indicates the spread of repetitive measurements, including measurements performed with the Lock-In amplifier. We prove the linearity of the cantilever response by measuring the cantilever amplitude a on resonance as a function of piezo drive amplitude. We find linear behaviour over the full range of accessible cantilever amplitudes (up to $a = 1.6\,\mu$m in a measurement performed in a SEM before glass cell assembly).

Temperature shift

We find that the cantilever eigenfrequency strongly depends on temperature. This is probably caused by the different thermal expansion of Au and SiN and the resulting

4.1 Atom chip with microcantilever

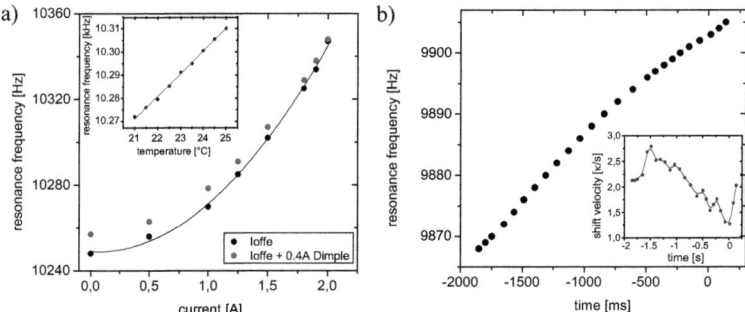

Figure 4.4.: (a) Measured change of the resonance frequency as a function of Ioffe wire current. We observe the quadratic dependence expected for resistive heating. The inset shows the temperature dependence of the eigenfrequency. We observe a frequency shift of 11.2±1.5Hz/K. (b) Measured temporal evolution of the resonance frequency during the experimental sequence. At $t = -1700$ ms the currents are set to a fixed value, at $t = 0$ ms the experiments are performed. The inset shows the velocity of the frequency shift calculated from the data.

change of tensile stress in the Au film. To calibrate the frequency shift we set the temperature of the watercooling system to various values and measure the resonance frequency after sufficient time for equilibration (see inset of figure 4.4 (a)). The observed temperature shift amounts to 11.2 ± 1.5 Hz/K, about one linewidth per degree Celsius. This shift becomes important as the wire currents that are used for the magnetic traps during the experiment dissipate up to 7 W and thereby heat the chip. We infer the amount of wire induced heating with a measurement of the resonance frequency as a function of the wire current. We vary the static current in the Ioffe wire and repeat the measurement with an additional current in the Dimple wire. Figure 4.4 (a) shows the measurement together with a quadratic fit as expected from resisitive heating $P_R = RI^2$. The total shift for $I_I = 2$ A amounts to 86 Hz, corresponding to a temperature change of 7.7 K. In this measurement the cantilever serves as a local thermometer on the chip.

During an experimental sequence where a Bose-Einstein condensate is produced and brought close to the cantilever for coupling, we observe a rapid drift of the resonance frequency due to the thermal drift (see figure 4.4 (b)). The drift can cause uncontrolled cantilever amplitudes when piezo excitation is applied, and we have to minimize it at the end of the sequence when the condensate is coupled to the cantilever ($t = 0$ ms in the figure). Therefore we design the sequence such that all currents are set to a fixed value at the beginning of the evaporation phase ($t = -1700$ ms), and in subsequent phases only the offset fields from the coils are changed. This imposes a constant heat load and allows the chip to approach steady state. The behaviour of the resonance frequency can be seen more clearly when looking at the velocity of the frequency shift over time (see inset in 4.4 (b)). After switching of the wire currents at $t = -1700$ ms, the velocity increases within 200 ms to a maximum value of ~ 3 κ/s, followed by a quite linear decrease. At the time of the coupling phase it has reduced to 1.3 κ/s. After the coupling phase, the currents are ramped down and the velocity increases again. Compared to a previous sequence, the shift velocity is reduced by a factor $2-3$. For shift velocities $\lesssim 1$ κ/s, the cantilever can reach its steady state amplitude within a time $\sim 3\tau = 330$ ms. This is fairly satisfied during the respective time before the coupling period, and when probing e.g. the cantilever resonance spectrum at $t = 0$ we observe a slightly broadened Lorentzian with $\kappa = 8.2$ Hz.

An additional effect comes from absorption of laser power from the readout laser. We observe a linear dependence up to optical powers of 0.8 mW and find a frequency shift of 129 Hz/mW. Also the light from the MOT changes the frequency by ~ 20 Hz.

When driving large currents and operating the readout laser simultaneously, the temperature of the cantilever can increase by up to 20 K. This could have implications for the desorption rate of Rb adsorbed on the cantilever surface (see chapters 2.3.2 and 5.4.3).

4.2. Vacuum, Lasersystem, Electronics

In the following I describe the peripheral experimental system.

4.2.1. Vacuum setup

The chip assembly described above is glued to a Pyrex cell[6] with an inner edge length of 30 mm and one open face. The outside of the cell is anti-reflection coated for 780 nm, and the open side has a bevel of 0.3 mm to restrict the glue meniscus. Before the chip is attached, we drill a 23 mm diameter hole in the opposite side and glue a glass-to-metal adaptor[7] to the cell. The two gluing steps that complete the optical cell are critical. Due to the different thermal expansion of the chip and the glass cell, substantial stress can build up and even cause a break of the cell. Furthermore, the glued bond has to sustain the vacuum bakeout process and seal the system at a base pressure of $\sim 10^{-10}$ mbar. Again we use Epo-Tek 353 ND after thorough outgassing due to its good UHV compatibility and mechanical stability.

An optimized curing schedule is used to minimize the emergence of stress during hardening of the epoxy [82]. After application of the glue we let it harden for two days. This allows the glue to shrink almost to the final volume while it is still

---
[6]Hellma 704.027-BF without label
[7]Caburn DN40CF Pyrex to stainless steel adaptor

4.2 Vacuum, Lasersystem, Electronics

viscous, and stress is avoided. For the final curing, the whole cell and glass-to-metal transition is heated with a slow ramp over 1.5 h to a temperature of 150 °C. After curing for 1 h at this temperature we slowly ramp down again. This procedure proved to be very reliable.

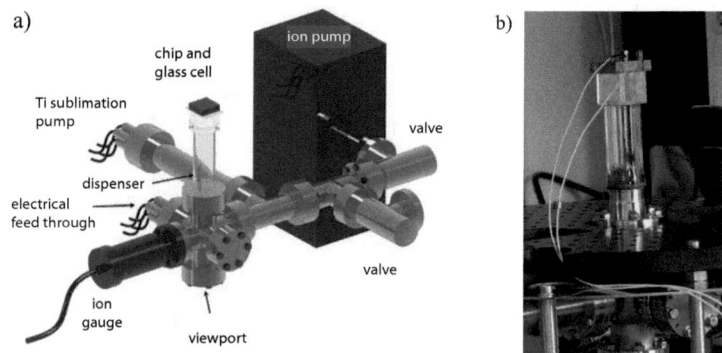

Figure 4.5.: (a) Schematic overview of the vacuum system; graph adapted from [82]. (b) Glass cell with chip and copper block mounted on a glass-to-metal adaptor and flanged to the vacuum system.

On the backside of the chip we place a copper block that includes a water cooling channel and an "U" shaped bar that provides the quadrupole field for the MOT [255]. It replaces bulky anti-Helmholtz coils and creates a field configuration close to an ideal quadrupole.

The glass-to-metal adaptor is flanged to a stainless steel six-way cross with DN40 CF flanges. It contains an ion gauge[8], a vacuum feedthrough, and three Rb dispensers[9] as source for thermal Rb atoms. A viewport on the bottom of the six-way cross provides optical access to the chip from below, and is e.g. used to monitor the performance of the MOT with a small video camera. The pressure in the system is maintained by an ion pump[10] and from time to time (every few months) by a Ti-sublimation pump. Figure 4.5 shows an overview of the vacuum system.

After an initial bakeout of the steel chamber at 150 °C, the entire system including the glass cell is baked at 100-110°C. The low temperature is necessary due to the integrated piezo and the glued glass cell. After a few weeks of bakeout we obtain

---
[8]Leybold Ionivac IE514 Extractor
[9]We have installed two dispensers based on Rb chromate (SAES Rb/NF/3.4/12 FT10+10) and one dispenser involving a metallic Rb/In alloy (Alvatec Alvasource AS-RbIn-5-F), but used only one of the SAES dispensers so far.
[10]Varian VacIon Plus StarCell 25 l/s

a pressure of 1×10^{-8} mbar at 110°C and 3×10^{-10} mbar after cool down to room temperature and a few runs of Ti sublimation.

4.2.2. Laser system

We use a laser system based on home made diode lasers. It provides the optical fields for laser cooling, optical pumping and absorption imaging. One additional laser is used for the optical readout of cantilever motion. The system is compact and fits on the area of 1.4×1.2 m^2. We cover it with a foam board box to reduce dust, acoustic vibrations and air circulation. To decouple the beam alignment at the vacuum cell from the laser system, all beams are coupled to polarization maintaining single-mode fibers[11].

For the manipulation and detection of atoms we have to derive four different frequencies of laser light close to the D2 line of ^{87}Rb at 780.24 nm. An overview of the involved transitions is shown in figure 4.6. We use three different lasers that typically run with Sharp diodes[12] with 120 mW nominal output power. Two of the lasers contain a grating in Littrow configuration [256] to reduce the linewidth. Their frequency is stabilized by Doppler-free saturation spectroscopy in a Rb vapour cell [257]. We generate a locking signal by frequency modulation of the laser at 110 MHz [258]. The feedback of the lock regulates the grating position with a piezo (integral path) and the laser current (proportional path), which results in a laser linewidth of ~ 300 kHz.

An additional laser is used for the independent readout of the cantilever. We operate it with a diode running at 830 nm to be far off resonant from optical transitions. This avoids scattering by the atoms, and at low intensity, dipole forces remain negligible. It is thus possible to continuously monitor the cantilever, also in the presence of atoms nearby.

A schematic overview of the system is shown in figure 4.7. The following beam lines constitute the setup:

Cooling To obtain sufficient power for laser cooling we use a master-slave configuration. The master laser is grating stabilized and frequency locked to the $F = 2 \rightarrow F' = (2,3)$ crossover resonance. We feed the light through a double-pass AOM to change the frequency shift without altering the beam alignment. For cooling we use a maximal red detuning of $\Delta = -11\gamma$ while for imaging in the presence of magnetic fields we use blue detuning up to $+3\gamma$, where $\gamma = \Gamma/2\pi = 6.065$ MHz the natural linewidth of the D2 line. The beam is then injected into a slave laser without grating. It provides the full output power of the diode while the frequency stability is inherited from the master laser. We typically run the slave with a cur-

---

[11] Thorlabs PM-780HP
[12] Sharp GH0781JA2C

4.2 Vacuum, Lasersystem, Electronics

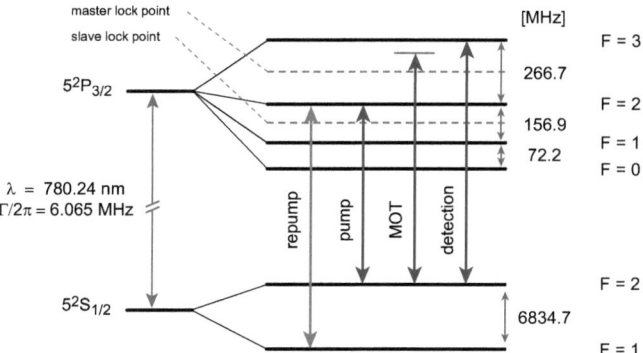

Figure 4.6.: Level scheme of the ^{87}Rb D2 line and used laser frequencies. The locking point of the two spectroscopy stabilized lasers is indicated.

rent of 120 mA and obtain up to 80 mW optical power after an optical isolator. A switching AOM is used to adjust the power and for fast switching. Finally, the light is split into four paths, each of which is coupled into a polarization maintaining single-mode fiber that guides it to the experiment. Two fibers provide the beams that are reflected on the chip under 45° and carry 15 mW each, and the remaining two provide horizontal beams carrying 5 mW each.

Imaging A small fraction of the slave light is split to a path with an additional switching AOM to provide light for absorption imaging. We use two axis for imaging to be able to cover both the cooling and the cantilever region. The power is chosen such that the peak intensity in the collimated beam after the fiber is a small fraction (< 20 %) of the saturation intensity $I_s = 1.67$ mW/cm^2 of the cycling transition $F = 2 \rightarrow F' = 3$.

Repumping The cooling light also drives the non-resonant transition $F = 2 \rightarrow F' = 2$ with a probability $\sim 1/2000$ for each scattering event. This populates the $F = 2$ excited state which can decay also to the $F = 1$ ground state. To bring population in this state back to the cooling cycle we use a grating stabilized repump laser. It is locked to the $F = 1 \rightarrow F' = (1, 2)$ crossover and shifted to resonance with a switching AOM that is also used to set the power level. We overlap the beam with the 45° cooling beam on a polarizing beam splitter. After the fibers we have a power of 5 mW each.

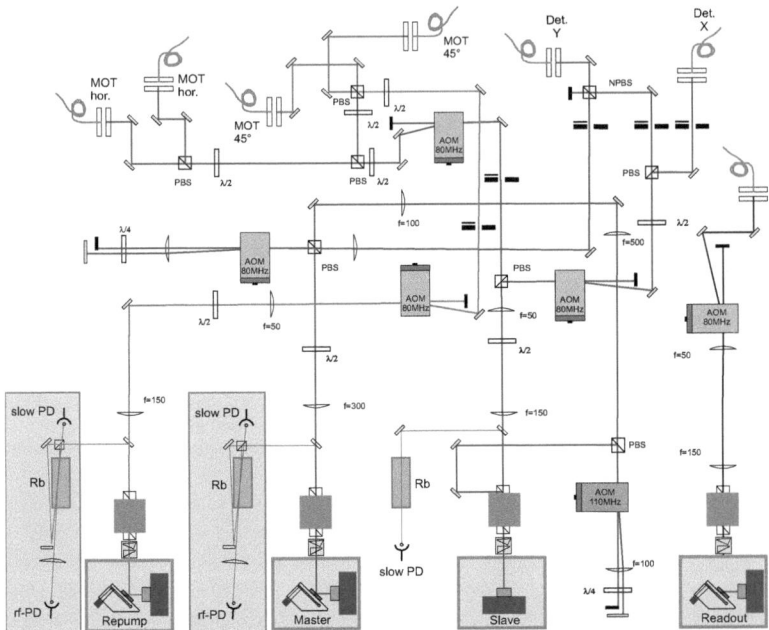

Figure 4.7.: Diode laser system used for cooling, optical pumping, imaging, and cantilever readout. All beams are controlled with AOMs and electromechanical shutters, and are coupled to polarization maintaining single-mode fibers.

Optical pumping Light for optical pumping is derived from the unshifted master laser with a typical power of 17 mW. The pumping light is frequency shifted to the $F = 2 \rightarrow F' = 2$ transition by an AOM in double-pass configuration. The light is overlapped with the imaging beam for the y-axis on a non-polarizing beam splitter and coupled in the same fiber.

Fast switching within less than one microsecond is accomplished by switching AOMs for each beam. However, residual scattering into all diffraction orders also without RF driving of the AOM crystal leaves some light in the fibers and thereby reduces the magnetic trap lifetime. Electro-mechanical shutters[13] with a transient time of 1 ms and a delay of $5 - 7$ ms are used to completely block all beams during magnetic trapping phases.

---

[13] Harting Solenoid 2810025-0614

4.2 Vacuum, Lasersystem, Electronics

4.2.3. Current sources and magnetic field coils

Current sources

Stable current sources play a key role for atom number stability after evaporative cooling, low heating rates, and ultimate position reproducibility. The challenge is that they have to achieve exceptional stability and low noise, while simultaneously providing fast switching.

source	use	current [A]	voltage [V]	rms noise [μA]	switching [μs]
Delta	copper U-bar	+100	70	10^4	-
Kepco BOP	transport coil	±20	20	750	150
FUG	B_y coil	+15	20	430	1000
Typ-Claus	Base-Chip MOT	+5	15 (60)	370	40
i-source	(2x) $B_{b,x}$, $B_{b,z}$ coil	±5	10 (10)	28	1500
i-source	2 mm Ioffe	±3	10 (9)	28	40
High Finesse	3 mm Ioffe	+3	12 (20)	140	50
High Finesse	Waveguide	+3	10 (17)	140	50
High Finesse	Shift Dimple	+1	30 (40)	125	100
i-source	Dimple	±1	10 (10)	28	150

Table 4.1.: Current sources used in the experiment. Bipolar sources are indicated by ± currents. Noise and switching time values are measured in the setup. All wires on the microtrap chip are connected, and only one of the respective sources is grounded while all others are floating.

Overall we have 11 current sources in operation, table 4.1 gives an overview. We quote the specified maximum current and output voltage and give the measured maximum open circuit voltage (in brackets) for sources used for wires. This is important for wire protection considerations. To avoid wire overloading we introduce safety resistors that limit the maximal possible current of all sources connected to a wire to an uncritical value. Additionally we use fuses that we test for proper breakdown current. The given rms noise and switching times are measured in the setup with the experiment control providing the set voltage. The loads are in general floating and are mainly ohmic for wires (typically $0.6 - 2.5$ Ω wire resistance plus $2 - 40$ Ω safety resistors) and inductive for coils (see below).

For the noisy sources we use switches[14] to detach them during critical phases. The sources called *i-source* were developed and built in our group [170]. They combine exceptional low noise (5×10^{-6}) and long term stability (1×10^{-5}) with fast switching

---
[14]We use Crydom solid state relais for unipolar sources and a network of mechanical relais for bipolar sources.

coil	windings #	field per current [G/A]	inductivity [mH]	resistance [Ω]
$B_{b,x}$	35	5.75	0.15	0.28
$B_{b,y}$	65	6.95	1.62	0.73
$B_{b,z}$	76	4.34	1.32	0.92
Transport	33	-/ 2.82 G/(cm A)	0.24	0.20

Table 4.2.: Coils for homogeneous and quadrupole magnetic fields. Values are quoted for pairs of coils connected in series.

($\sim 15\ \mu$s for an ohmic load of 1 Ω). The *Typ-Claus* source is an old version developed earlier in the group. All other sources are commercially available.

Magnetic field coils

In principle, all necessary magnetic fields could be generated on the chip. However, the fields from wires on the chip have strong gradients and it is advantageous to have homogeneous fields available. We use a cage with three pairs of coils in Helmholtz configuration to provide homogeneous fields along all axes. Additional windings on the x-axis connected in anti-Helmholtz configuration provide a strong quadrupole field for a magnetic transport in a waveguide. To compensate for the earth magnetic field we add small bias fields rather than using additional coils.

The mechanical design is adapted from [150] and modified with respect to easier machining and the different chip geometry. We have implemented water cooling for each coil to reduce thermal drifts. We use Kapton isolated copper wires of 1.22 mm diameter, and spooling is done with a specialized machine at the MPI of plasma physics (IPP) in Garching. The strongest field can be generated along the y-axis with a maximum field of 100 G.

4.3. Experimental sequence for BEC preparation

The preparation of a Bose-Einstein condensate is accomplished by a series of individual steps, each of which has to be optimized carefully. We control all the relevant parameters of the experiment by a computer with several digital and analog output cards[15]. In total we have 20 analog 16 bit channels, 32 analog 13 bit channels and 48 digital channels. We use a time-sequencing program with a time resolution set to 50 μs which was initially written by Jakob Reichel and modified by Pascal Böhi. For the acquisition and analysis of absorption images we use an additional computer with a MATLAB-based application written by Pascal Böhi.

---

[15]National Instruments PCI-6733, PCI-6723, and PCIe-6259

4.3 Experimental sequence for BEC preparation

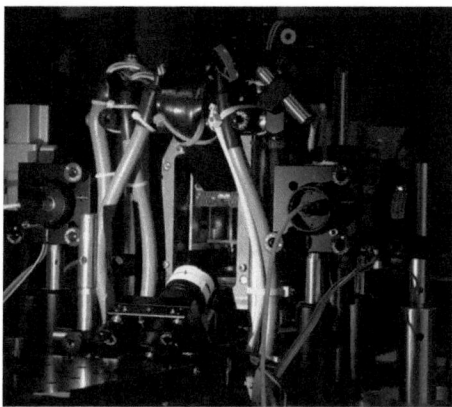

Figure 4.8.: Vacuum cell surrounded by magnetic field coils, MOT and imaging optics.

Mirror MOT

The first step of the experiment is the preparation of a cloud of atoms with sufficiently low temperature to allow magnetic trapping. This is realized with a Magneto-Optical-Trap (MOT) [259] that combines laser-cooling to extract energy from thermal atoms, and magnetic level shifts that introduce a position dependence to the cooling forces, thereby creating a trap minimum where the atoms are collected.

The cooling principle is based on the Doppler shift of the resonance frequency of a moving atom [260, 261, 262]. An atom moving in a pair of red detuned laser beams will absorb light predominantly from the beam opposing its movement. The associated scattering of photons imparts a net momentum on the atom which slows it down. With beam pairs along three orthogonal axes, this provides cooling of a gas. In our case, the chip obstructs one half-space for optical access, and we use the configuration of a mirror-MOT [252, 16]. It replaces two beams by introducing a mirror which is applied on the chip surface (see chapter 4.1). The four required beams are circularly polarized and red detuned by a few linewidths from the cycling transition $F = 2 \rightarrow F' = 3$.

All light is led to the experiment via polarization maintaining single mode fibers. After the fibers, the cooling beams are collimated by $f = 60$ mm achromatic lenses to an $1/e^2$ diameter of 12 mm. Two of the beams are aligned under grazing incidence with the chip, parallel to the base of the "U" along the x-axis, and the remaining two beams, which also include the repumping light, are inclined under an angle of 45° with respect to the chip surface along the y-axis. The imaging beams need a smaller area and are collimated to a diameter of 8 mm. Figure 4.9 shows an overview on

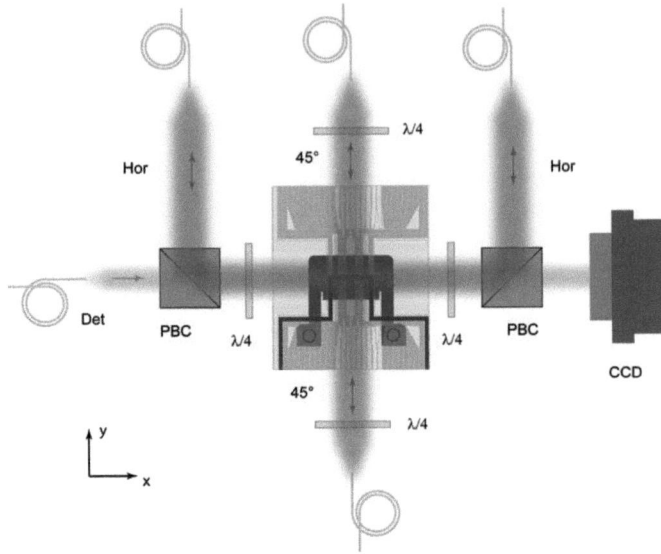

Figure 4.9.: Laser beams and wire structures for the operation of a mirror MOT and optical molasses. Currents in a copper U-bar (brown) and a wire on the base chip (red) together with a homogeneous field $\boldsymbol{B}_0 = (B_{b,y}, B_{b,z})$ provide a 3D quadrupole field. Two horizontal beams and two beams inclined by 45° with respect to the chip provide cooling and repumping light. Imaging light along the x-axis is combined with (and split from) the cooling light by polarizing beam splitters (PBS), while circular polarization is set by quarter wave plates ($\lambda/4$). Imaging along the y-axis is not shown here.

the various beams at the vacuum cell.

The atomic species ^{87}Rb is provided by resistively heated dispensers that set free a constant flux of thermal atoms. Operation of the dispensers close to their threshold (in our case with a constant current of 3.6 A) increases the vacuum pressure slightly, and the pressure gauge typically shows $\sim 6 \times 10^{-10}$ mbar. We perform three different stages of the MOT:

1. Loading phase This MOT phase is used to load atoms from the Rb background gas. We operate it at a current of 55 A in the U-bar and offset fields $B_{b,y} = 10$ G, $B_{b,z} = 4$ G. This defines the trap center at a distance of 7.2 mm from the U-bar, and about 5.2 mm from the mirror on the chip surface. The resulting gradient amounts to 15.2 G/cm along the strongest direction. We find an optimal detuning of $\Delta = -2.3\,\gamma$ and typically load 1.5×10^7 atoms in 6 s.

4.3 Experimental sequence for BEC preparation

2. Chip MOT For this phase we change from the U-bar as field source to a U-shaped wire on the base chip. During field switching the lasers are briefly turned off. The field is now generated by a current of 4.6 A and a homogeneous field of $B_{b,y} = 2.3$ G. The trap center is then ramped within 20 ms to the chip surface by reducing the current to 2.25 A. This also increases the gradient and compresses the cloud. The steady state atom number in this stage would be much lower than in the previous phase, and we choose a short duration of 5 ms. The position of the cloud is optimized for loading into the magnetic trap in a later step.

3. Compressed MOT To obtain maximal density and minimal temperature we increase the detuning to $\Delta = -10\gamma$ and reduce the repumping power for a duration of 3 ms. This reduces the scattering rate, while each scattering event extracts more energy, and heating by reabsorption of scattered light is suppressed. This leaves us with about 90 % of the atoms collected in the first phase at a temperature of 80 μK.

Molasses

To reduce the temperature below the achievable values in a MOT, we perform optical Molasses cooling. This involves carefully nulled magnetic fields which we achieve by Hanle spectroscopy [263, 264] and reduced cooling and repumping power. The detuning is set to $\Delta = -11\gamma$ and we find an optimal duration of 2.8 ms. This phase reduces the temperature below 10 μK.

Optical pumping

After laser cooling, the atoms are distributed over all sublevels of the ground state. For magnetic trapping we have to prepare the ensemble in a spin polarized, weak-field seeking state, in our case the state $|F = 2, m_F = 2\rangle$. To achieve this we irradiate repumping light to ensure only $F = 2$ population, and circularly polarized pumping light at the $F = 2 \rightarrow F' = 2$ transition along the y-axis to polarize the sample. To define the quantization axis we set a homogeneous field of 2 G along y. This completes the laser manipulation of the atoms, and after a pumping duration of 100 μs we obtain 1.2×10^7 atoms at a temperature of 10 μK.

Magnetic traps and transport

We load up to 8×10^6 atoms into a Ioffe trap created by a 3 mm long, z-shaped Ioffe wire[16]. We operate it with a current of 3 A and a homogeneous field of $B_{b,y} = 9.8$ G which defines the trap-wire distance $z_{t,0} = 600$ μm. The overlap between the cloud after laser cooling and the magnetic trap is optimized by the magnetic fields during the last MOT phase.

---

[16]We measure the atom number after a hold time of 0.5 s to allow for equilibration

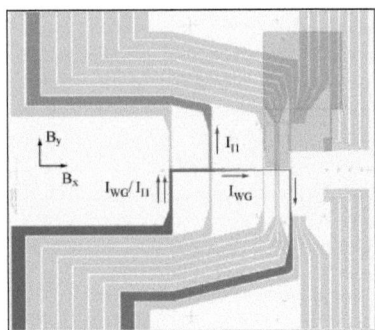

Figure 4.10.: Wire layout for the first magnetic trap (dark grey, I_{I1}) and the wire guide for the transport (medium grey, I_{WG}).

For experiments with the cantilever we have to transport the atoms over a distance of 6.4 mm along the x-axis. This is realized by ramping to a wire guide with a superimposed quardupole field along the x-axis from an anti-Helmholtz coil pair to provide axial confinement. Applying an additional homogeneous field along x enables to shift the minimum of the quadrupole [150]. The quadrupole provides an axial gradient of 22.6 G/cm, while the wire guide generates a transverse gradient of 250 G/cm for the used wire current $I_{WG} = 2$ A and offset field $B_{b,y} = 11.5$ G. The large difference in the gradients ensures that the superposition of the transverse component of the external quadrupole field with the waveguide field leads only to a small distortion of the transverse trapping potential. We find that the optimal atom-surface distance for the transport changes from $z_{t,0} = 400$ μm at the beginning to $z_{t,0} = 370$ μm at the position of the cantilever. We find an optimum transport ramp duration of 170 ms and transfer typically 50% of the atoms to an Ioffe trap at the cantilever (MTrap4, see table 4.3).

RF Cooling and condensation

To reduce the temperature and increase the phase space density of the cloud to achieve Bose-Einstein condensation we use radio-frequency (RF) evaporative cooling [147]. Evaporative cooling is based on the fact that a selective removal of the most energetic atoms leaves the remaining part of the cloud at a lower temperature after thermalization. Energetic atoms can be removed by a controlled opening of the trap at a truncation energy $E = \eta k_B T$ with an optimal $\eta \approx 6$ by a RF field. The field induces spin-flip transitions to untrapped magnetic sublevels, and the frequency selects the energy at which the transitions occur.

4.3 Experimental sequence for BEC preparation

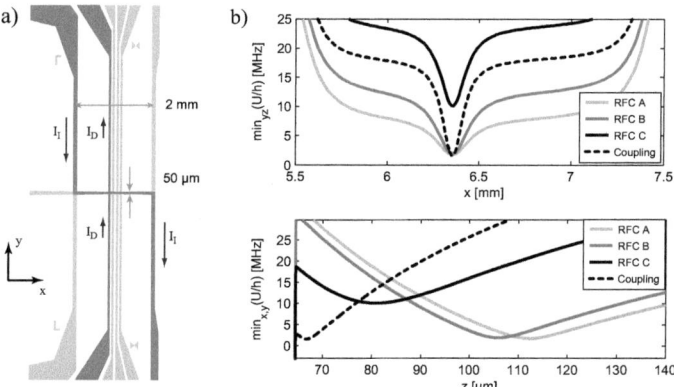

Figure 4.11.: (a) Wire layout of the 2 mm Ioffe trap. The used Ioffe (I_I) and Dimple (I_D) wires are depicted in dark grey. (b) Cuts through the trapping potential along the trap minimum in x (top) and z (bottom) direction. Shown are the three traps for evaporative cooling and a trap close to the cantilever for coupling. The cantilever surface is located at $z = 64$ μm and causes a reduced trap depth.

We use two frequency generators[17] followed by a +18 dB amplifier and a switch, which drive a self-made coil placed close to the cell to generate the RF field.

Rapid evaporation requires a high collision rate for thermalization, while the inelastic collision rate should remain small at the same time. Due to the different density dependence of the two rates (see chapter 2.5.1) it is advantageous to begin the evaporation in a trap with the highest possible trap frequency. At the end of the cooling phase, when the density has increased by several orders of magnitude, three-body collisional loss becomes severe and the trap frequency has to be reduced to a moderate value.

We use a 2 mm wide, z-shaped wire with a crossing dimple wire to create a Ioffe-type Dimple trap at this stage. Figure 4.11 (a) shows the relevant part of the wire layout.

In our experiment, we have to respect two untypical boundary conditions during evaporative cooling. First, we want to prepare the condensate as close as possible to the surface to avoid large transport distances to the cantilever. Yet, we have to ensure that the distance d between the trap and the cantilever surface is sufficiently large for each stage to avoid uncontrolled surface loss. Second, we want to avoid changes in the wire currents to minimize changes in the thermal load of the chip and

---
[17]We use a SRS DS345 for the first and a HP E4431B ESG-D with analog FM for arbitrary frequency ramps during the final two stages

the resulting frequency shift of the cantilever (see chapter 4.1.2). This influences the choice of the phases and trap parameters.

We use three stages of evaporative cooling. The first two stages are performed in traps at a relatively large distance $d > 40$ μm, while for the third stage we approach the cantilever to $d = 16.5$ μm. We observe a lifetime of 2.5 s after cooling close to the transition temperature in this trap, only slightly reduced from the value of 3.2 s far away from the surface. Table 4.3 gives an overview of the parameters for the cooling phases.

Trap	Δt [ms]	I_I [A]	I_D	$B_{b,x}$ [G]	$B_{b,y}$	f_z [kHz]	f_x	$z_{t,0}$ [μm]	B_0 [G]	ν_{RF0} [MHz]	ν_{RF1}
MTrap4	20	1.855	0.25	2.3	10	0.31	0.06	355.6	1.7	-	-
RFC A	700	1.855	0.25	5.5	35	3.64	0.32	112.9	1.1	25	8
RFC B	600	1.855	0.4	9.0	37	3.86	0.43	105.9	1.3	8.5	2.5
RFC C	500	1.855	0.4	18.0	47	2.70	0.62	80.8	7.1	7.2	4.98
MTrap8	1-20	1.855	0.4	14.5	58.3	10.02	0.81	66.8	1.0	-	-
Detection	3	0.6	0.05	4	10.5	0.75	0.14	119	2.1	-	-

Table 4.3.: Magnetic traps at the cantilever after transport (MTrap4), for RF Cooling (RFC A-C), and for coupling (MTrap8). We quote the duration of the phase Δt, Ioffe wire current I_I, Dimple wire current I_D, axial (f_x) and transversal ($f_z \approx f_y$) trap frequency, trap-wire distance $z_{t,0}$, magnetic field in the trap minimum B_0, and start (ν_{RF0}) and stop (ν_{RF1}) frequency of the RF sweep.

The first stage (RFC A) uses a trap with a shallow dimple of only ~ 4 MHz. This has mainly technical reasons: We find a better performance for the evaporation ramp when it spans only the part of the trap above the Dimple. Because the RF-generator used for the subsequent ramps spans only 8.5 MHz we choose the depth of the Dimple smaller than this value.

For the second stage (RFC B) we transform the trap such that the remaining cloud is contained entirely in a 8 MHz deep dimple with a mean frequency of $\omega_{ho}/2\pi = 1.84$ kHz, providing a high collision rate.

For the third stage (RFC C) we have to relax the trap to reduce the inelastic collision rate. Furthermore we choose the minimal distance from the surface where surface effects do not yet contribute (see also chapter 4.4.1).

Figure 4.11 (b) shows the resulting potentials along the x and the z axis for the three cooling stages and for a typical trap used for measurements close to the cantilever.

We produce pure BECs of typically 2000 atoms after an overall evaporation time of 1.8 s. Figure 4.12 shows the atom number and optical density (OD) together with three examples of absorption images during the final part of the third RF ramp. The increasing OD for falling atom number is a sign for runaway evaporation [147]. The

4.3 Experimental sequence for BEC preparation

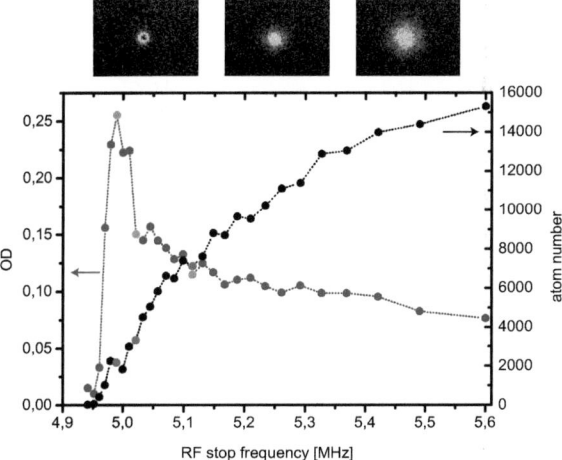

Figure 4.12.: RF cooling across the phase transition. Thermal cloud (top right, $\nu_{RF1} = 5.11$ MHz), cloud at the transition temperature (top middle, $\nu_{RF1} = 5.02$ MHz) and fully condensed cloud (top left, $\nu_{RF1} = 4.99$ MHz) after 8 ms time of flight (corresponding datapoints are indicated in orange / grey). Bottom: Optical density (OD) and atom number as a function of RF stop frequency of the last cooling stage (RFC).

abrupt rise for frequencies $\nu_{RF1} < 5.02$ MHz indicates the phase transition to BEC. For $\nu_{RF1} < 4.95$ MHz all atoms are removed from the trap, corresponding to the trap bottom at a magnetic field of $\min(B) = 6.94$ G. In our experiments, we typically use a value of $\nu_{RF1} = 4.98$ MHz to prepare BECs without discernible thermal cloud.

Detection

We detect the atoms by the standard technique of absorption imaging [6]. We irradiate the atoms with a 50 µs long pulse of σ^+ polarized light resonant with the cycling transition $F = 2 \rightarrow F' = 3$ and observe the shadow the atoms cast on the beam with a CCD camera. To discern the spatial modulations on the imaging beam from the atomic signal, two images are taken, one with atoms (A) and a second without (B). To take out camera dark counts and stray light, a third image (D) without imaging beam is taken typically once a day. The three images allow to pixelwise calculate the optical density

$$\mathrm{OD} = -\ln \frac{A-D}{B-D}. \tag{4.2}$$

For an imaging intensity much smaller than the saturation intensity I_s, this links directly to the atomic column density $\tilde{n}(x,y) = \mathrm{OD}(x,y)/\sigma_0$ with the scattering cross section $\sigma_0 = 3\lambda^2/2\pi$. The atom number located in an area F is then calculated by summing over all relevant pixels,

$$N = \frac{F}{\sigma_0} \sum_{i,j} \mathrm{OD}(i,j). \tag{4.3}$$

For the initial phases before the transport we image along the y-axis and use a video camera from JAI[18]. For the traps after the transport we image along the x-axis and use a camera from Apogee[19] with a quantum efficiency of 35% at 780 nm. The two imaging beams are provided by two fiber outcouplers, each with a $f = 40$ mm achromatic lens that collimates the beam to a diameter of 8 mm. We use a peak intensity of $\lesssim 0.2\,I_s$ to avoid saturation of the atoms. To image the shadow on the CCD camera we use two achromatic plano-concave lenses back to back, such that the imaging error is minimized. For the x-axis we use lenses with $f_1 = 100$ mm and $f_2 = 200$ m, resulting in a magnification of 2.4 and a resolution of 9 μm. We observe an imaging noise of 12 atoms rms on the respective area.

4.4. Atoms close to the surface

For the experiments presented in chapter 5 we want to use BECs in tight traps at very small atom-surface distance. In chapter 2 we have discussed various effects that influence the atoms under such conditions. In the following we characterize the trap lifetime and heating rates. We find that three-body collisional loss and technical current noise are severely limiting effects in the tight traps we use for the experiments in chapter 5. Furthermore, Johnson-Nyquist noise and surface evaporation contribute to a lifetime reduction independent of trap frequency.

4.4.1. Trap lifetime vs. atom-surface distance

First we determine the influence of Johnson-Nyquist noise and surface evaporation on the trap lifetime close to the metallized side of the cantilever. For this measurement we prepare thermal clouds close to the transition temperature at $T = 1.5\,T_c$ and measure the atom number decay for various atom-surface distances. We use a

---
[18]CV-M50-IR, 752 × 582 pixels, 8.4 μm pixel size, 8 bit dynamics.
[19]Ap1E with Kodak sensor (KAF-0401E), 768 × 512 pixels, 9 × 9 μm^2 pixel size, 14 bit dynamics.

4.4 Atoms close to the surface

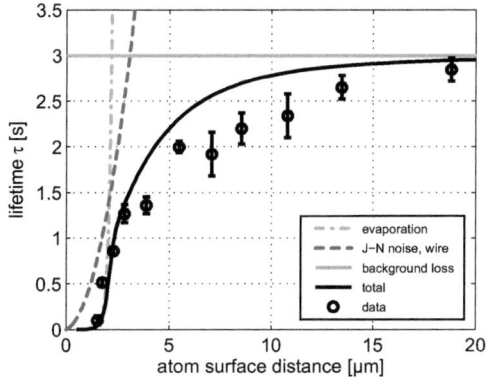

Figure 4.13.: Trap lifetime of a thermal cloud at $T = 1.5\,T_c$ as a function of atom surface distance. Initial atom number is $N(0) = 5000$ in a trap with $(\omega_x, \omega_y \approx \omega_z) = 2\pi \times (0.5, 5.2)$ kHz. Comparison with a calculation of atom loss due to surface evaporation (dash-dotted line) and Johnson-Nyquist noise (dashed) shows good agreement.

thermal cloud to reduce the density and thus avoid excessive three-body collisional loss.

We find that for distances $d > 15$ μm the lifetime of the atoms remains essentially unaffected by the surface, while for $d < 5$ μm it has reduced to half of the value for background loss, and falls off rapidly for $d < 3$ μm. Figure 4.13 shows the measurement together with a calculation of the relevant loss rates according to Eqs. 2.92 and 2.80. The observed lifetimes can be well described by surface evaporation and Johnson-Nyquist noise induced spin-flip loss.

The measurement shows that it is possible to perform measurements on a second timescale at a distance of only a few microns away from the surface. Note that for a BEC in front of a dielectric surface one expects no lifetime limitation due to the surface as long as there is no heating.

4.4.2. BEC lifetime vs. trap frequency

Three-body collisional loss is a major limitation for the lifetime of a BEC at high trap frequency (see chapter 2.5.1). The second effect with strong trap frequency dependence is technical current noise induced heating (see chapter 2.5.3). In traps close to the surface, where surface forces reduce the trap depth, heating also translates into trap loss via evaporation or sudden loss (see chapters 2.4.2). Both loss mechanisms give rise to non-exponential atom loss.

To quantify the collisional loss we prepare pure BECs and perform measurements of the trap population as a function of hold time for several trap frequencies at both small and large atom-surface distance. For simplicity we fit the data with exponential decay laws to obtain the trap lifetime.

To investigate heating rates we do not rely on temperature measurements, as they bear large errors for the small clouds in our experiments. We quantify the lifetime of the condensates by visually inspecting the absorption images and assessing the behaviour of the measured cloud radii. This allows us to infer the hold time after which the condensate is heated above the transition temperature to a completely thermal cloud. This is referred to as the BEC lifetime.

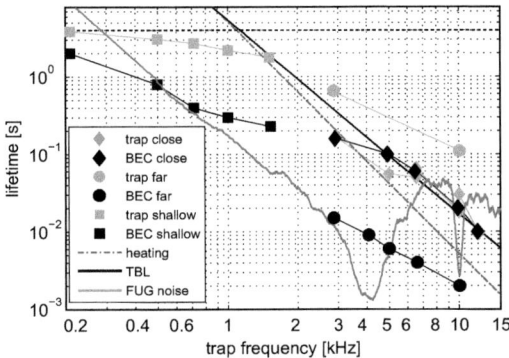

Figure 4.14.: Atom number (grey) and BEC lifetime (black) measured as a function of trap frequency for three different trap configurations. Diamonds (close): close to the surface as used for coupling experiments (see chapter 5.7). Circles (far): Similar trap configuration as before, but far from the surface. Squares (shallow): Traps with strongly reduced field gradients due to smaller I_I and $B_{b,y}$. The solid black line shows a calculation of the lifetime for three-body collisional loss for a cloud with $N_c = 2000$ atoms in a trap with an aspect ratio $\omega_x/\omega_z = 0.1$. The grey lines are calculations of the current noise induced thermalization rate either using the rms value (dash-dotted dark grey, "heating") for the used sources (FUG, i-source 2A) or the measured noise spectrum of the FUG source (grey, "FUG noise"). The dotted black line shows the background lifetime.

Figure 4.14 shows the measurements together with a calculation of the collisional loss rate (Eq. 2.87) and calculations of the expected thermalization time $t_{th} = T_c/\dot{T}[S_I(\omega)]$ according to Eq. 2.99. We either use the measured rms noise of the used current sources (i-source 1 A and FUG, see table 4.1), or the noise power

4.4 Atoms close to the surface

spectral density $S_I(\omega)$ to calculate the heating rate $\dot{T}$.

We find that the trap lifetime changes by two and the BEC lifetime by three orders of magnitude over the studied range of trap frequencies. Unexpectedly, the BEC lifetime close to the surface is larger than far away, while the trap lifetime shows opposite behaviour. We attribute this to a shielding or evaporative cooling effect of the surface, which removes energetic atoms that remain trapped after RF cooling and that could otherwise heat the condensate by collisions. This is also consistent with the observation of the reduced trap lifetime close to the surface. For the traps with $f_z = 3 - 13$ kHz which we use for coupling measurements we do not observe thermalization, and the BEC lifetime is set by the trap lifetime to $\tau = 1 - 20$ ms. The observed values are consistent with three-body collisional loss, which thus gives a fundamental restriction for coupling experiments.

4.4.3. Trap frequency measurements and trap simulation

For the coupling to cantilever motion, the trap frequency is an important parameter. A first value can be obtained by a simulation of the trapping potential. We use a MATLAB skript written by P. Treutlein [82] to calculate the magnetic fields generated by measured currents in the wires on the chip and the coils. Our simulation takes into account the finite width and length of the wires as well as the rectangular geometry of the three pairs of coils. However, uncertainties originating from the position and the spooling of the coils lead to an error in the simulated fields on the order of 10%. Measurements of the trap frequency in the actual traps used for coupling are thus necessary. This allows us to adjust the homogeneous offset fields in the simulation to match the measured data, thereby achieving an agreement of better than ±3% (see below).

We employ trap modulation spectroscopy and use either the excitation of the c.o.m. mode or trap loss as indicator. The trap is modulated by a small modulation of the Ioffe current ($\Delta I_I/I_I$ typically a few times 10^{-3}), leading to a modulation of the trap minimum position along the z−axis with an amplitude of a few nanometer. This excites center of mass oscillations when the driving frequency is resonant with the trap frequency. For traps close to the cantilever, the modulation is thus similar to the action of the cantilever on the trap position (see chapter 5.1).

We also use trap loss as signature in these traps. In this case, the observed loss resonances have a relatively large width of 70−2000 Hz, indicating trap anharmonicity. The measured trap frequency is thus not $\omega_{z,0}$, the value corresponding to the undisturbed trap, but corresponds to the inverse of the oscillation period for rather large amplitude oscillations in a trap that is deformed by surface forces (see chapters 2.4.1 and 5.1). As this is the relevant frequency for the coupling measurements, we refer to it as ω_z. Figure 4.15 shows such measurements and respective simulations for comparison. The relative uncertainty in the obtained ω_z is $< \pm 3$ %, which is also the amount of anharmonicity (for a cloud oscillating up to the barrier, a simulation

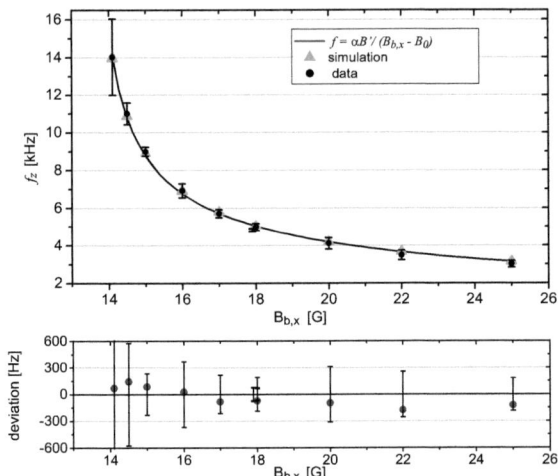

Figure 4.15.: Top: Measured (black points) and simulated (grey triangles) trap frequency for several traps at a few micron distance from the cantilever as used for dynamical coupling measurements in chapter 5.7. The errorbars give the width of the loss resonances used as signature for resonance. Solid line: Fit of Eq. 2.11 to the data with B' and B_0 as free parameters. Bottom: Deviation between simulation and measurement. The errorbars show the resonance width for comparison. The offset fields $B_{b,x}$ and $B_{b,y}$ in the simulation are adjusted to minimize the deviation from the data.

yields $(\omega_z(0) - \omega_z(b))/\omega_z(0) = 5\%$ anharmonicity). The simulation of the trapping potential deviates from the measured frequencies by less than $^{+1}_{-5}\%$ after adjusting the bias fields $B_{b,x}$ and $B_{b,y}$ once. We also compare the measured trap frequencies to the analytic prediction for an ideal Ioffe-Pritchard trap (Eq. 2.11).

5. BEC-resonator coupling via surface forces

This chapter covers the main results of this thesis. We show that atom-surface forces, which represent the most fundamental interaction of atoms with solid objects, can be harnessed to couple the motion of mechanical oscillators to the motion of trapped atoms.

In the first theoretical section I introduce the principle of the coupling mechanism and derive analytical expressions for the strength of the coupling.

I then describe experiments, where we first characterize and analyze the surface potential in the distance range that is useful for dynamical coupling experiments. The focus is then put on experiments where the driven motion of the microcantilever is coupled to collective atomic motion. We study in detail the properties of the coupling like the spatial range, overall strength, and spectral characteristics. We observe resonant excitation of individual collective modes of the BEC. The narrow resonances permit to control the coupling efficiently.

In the last section I present numerical simulations which we use to model the cloud dynamics.

5.1. Coupling via surface forces

In chapter 2.4.1 we have discussed the effect of the surface potential on a magnetic trap in vicinity of the surface. The main results were that the attractive potential leads to a reduced trap depth U_0, a shift of the trap minimum position z_t, and a shift of the trap frequency ω_z. We now consider the situation of a trap close to the tip of a cantilever that oscillates in its fundamental out of plane flexural mode with eigenfrequency ω_m.

In this setting, the position of the surface z_c and thus the surface potential U_s at the position of the trap are time-dependent, leading to a modulation of U_0, z_t, and ω_z at the cantilever frequency. Figure 5.1 shows a calculation of the potential modulation for typical experimental parameters. When the modulation is resonant with a collective mode of the atoms in the trap, coherent motion is excited and energy is transferred from the mechanical oscillator to the atomic cloud. This represents the general system of two coupled harmonic oscillators.

In the following we discuss the classical excitation of the two strongest modes of

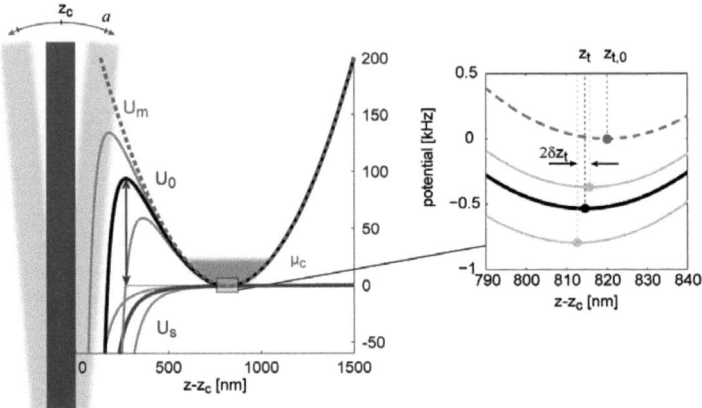

Figure 5.1.: The surface potential U_s of an oscillating cantilever modulates the trap minimum position and the trap frequency of the combined potential $U = U_m + U_s$. In the resonant case this excites collective modes of the atoms. The calculation shows the experimental situation of a trap with $\omega_{z,0}/2\pi = 10.0$ kHz at distance $d = z_{t,0} - z_c = 820$ nm from the static position of the cantilever in the presence of the Casimir-Polder potential of a perfect conductor. The static deformation of the trap amounts to $z_{t,0} - z_t = 5.4$ nm, $\omega_{z,0} - \omega_z = 104$ Hz, and $U_0 = 94$ kHz. Cantilever oscillations with amplitude $a = 80$ nm lead to a modulation of the barrier $\delta U_0 = 37.5$ kHz, of the minimum position $\delta z_t = 1.5$ nm, and of the trap frequency $\delta \omega_z = 110$ Hz.

the system, the center of mass (c.o.m.) or dipole mode, and the breathing mode. Note that for our trap geometry, both modes have the same resonance frequency for the condensed and the thermal component of the cloud.

5.1.1. Excitation of the dipole mode

The oscillation amplitude of the trap minimum position δz_t is given by the modulation of the surface force at the position of the trap center, and for a given cantilever amplitude a it amounts to

$$\delta z_t \cong \frac{1}{m\omega_z^2} \frac{\partial^2 U_s}{\partial z^2} a \equiv \epsilon a. \tag{5.1}$$

The coupling strength parameter ϵ has been introduced to describe the linear dependence on a for small amplitudes. This is valid as long as $\partial^2 U_s/\partial z^2$ and hence ω_z

5.1 Coupling via surface forces

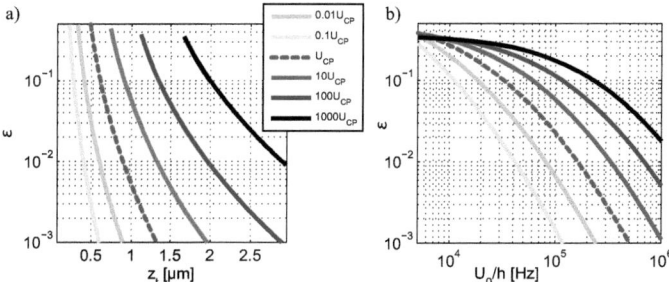

Figure 5.2.: Coupling strength parameter ϵ for different surface potential strength as a function of atom-surface distance (a) and barrier height (b). Chosen parameters are $\omega_{z,0} = 2\pi \times 10$ kHz and $U_{CP} = C_4/(z - z_c)^4$ with C_4 the CP-coefficient for a perfect conductor. Stronger surface potential can arise e.g. from adsorbates or surface contaminants, weaker potentials arise e.g. for molecular scale oscillators. The useful range for coupling experiments covers distance d $d \approx 0.4 - 4$ μm and trap depths $U_0 \approx (1 - 100)\hbar\omega_{z,0}$.

is approximately constant over one oscillation period.

An analytical calculation of ϵ as a function of atom-surface distance d and as a function of the barrier height U_0 is shown in Fig. 5.2 for a trap with $\omega_{z,0} = 2\pi \times 10$ kHz. Large values of ϵ up to $\epsilon = 0.3$ are found for small atom-surface distance, where the remaining trap is shallow and contains only one or two bound energy levels. A stronger surface potential enables large ϵ also for deeper traps. The atomic lifetime in such traps set the timescale of possible coupling experiments. A certain minimum coupling strength is thus necessary to achieve observable interaction, and for cantilever amplitudes $a \ll d$, an order of magnitude estimate demands $\epsilon \gtrsim 10^{-3}$. Depending on the strength of the surface potential, this restricts experiments to trap depths $U_0 \approx (1 - 100)\hbar\omega_{z,0}$, and for the trap frequency as in the given example, to possible atom-surface distances $d \approx 0.4 - 4$ μm.

It is well known, that an atomic cloud in a harmonic potential subject to a force modulated at the trap frequency is excited to center of mass (c.o.m.) oscillations [265, 101]. The differential equation for the center of mass coordinate Z in the presence of damping and driving can be written as

$$\ddot{Z} + 2\gamma\dot{Z} + \omega_z^2 Z = \delta z_t \omega_z^2 \cos(\omega_m t). \tag{5.2}$$

Here we have neglected the modulation of ω_z and the anharmonicity introduced by U_s, and we have quoted the force in terms of the shift of the trap minimum position. When damping is absent and the excitation is exactly on resonance $\omega_m = \omega_z$, the solution of this equation predicts a c.o.m. oscillation $Z(t) = b(t)\cos(\omega_z t + \phi)$ with a linear rise of the amplitude

$$b(t) = \frac{1}{2}\delta z_t \omega_z t. \tag{5.3}$$

In the presence of damping, the amplitude will saturate at a steady state value

$$b_{\max} \cong \frac{\delta z_t \omega_z}{2\sqrt{\Delta^2 + \gamma^2}} \xrightarrow{\Delta=0} \delta z_t \frac{\omega_z}{2\gamma}, \tag{5.4}$$

yielding a Lorentzian response as a function of the detuning $\Delta = \omega_z - \omega_m$, with the amplitude damping rate γ being the Half Width at Half Maximum (HWHM) of the resonance. Note that c.o.m. motion is essentially undamped, and dissipation may only arise from collisions with atoms from the background gas. But as their mean momentum is very large, this contributes to trap loss rather than amplitude decay of the oscillation. In contrast, the lifetime of the atoms in the trap will limit the coupling time t_h and thereby $b_{\max}$. This will lead to a Fourier limited resonance width of FWHM $= 1/t_h$.

Absence of anharmonicity is necessary both to allow an unbounded, linear increase in amplitude, and to leave the spatial distribution of the cloud unaffected by the c.o.m. oscillation. The decoupling of the internal cloud dynamics from the c.o.m. dynamics for harmonic potentials holds true also in the case of (strong) inter-particle interactions, and is known as the generalization of the Kohn theorem [113, 103, 115]. The situation changes significantly when the anharmonicity of the combined potential is taken into account. As will be discussed in more detail in Section 5.8, it leads to excitation of higher modes, dephasing, and a nonlinear rise of the amplitude.

5.1.2. Parametric excitation

The modulation of the trap frequency $\delta\omega_z$ provides a parametric excitation and leads as well to coupling. Its magnitude is quantified by

$$\delta\omega_z \cong \sqrt{\frac{1}{m}\frac{\partial^3 U_s}{\partial z^3}a} \equiv \xi\sqrt{a}, \tag{5.5}$$

where the proportionality factor ξ describes the dependence on a for small amplitudes. In the distance range found above, the relative frequency shift amounts to $\delta\omega_z/\omega_z \approx 10^{-3} - 10^{-1}$.

For a single particle there is an infinite series of parametric modes [265, 266] which are excited by modulation of one of the parameters of the oscillator, e.g. the trap frequency. The differential equation describing the situation is the damped Mathieu equation

$$\ddot{z} + 2\gamma\dot{z} + [1 + q\cos(\omega t)]\omega_z^2 z = 0. \tag{5.6}$$

5.1 Coupling via surface forces

Here, γ is the amplitude damping rate of the atomic motion and q describes the relative amplitude of the modulation of the spring constant. This equation shows instabilities, so called parametric resonances, for excitation frequencies

$$\omega_n = \frac{2\omega_z}{n} \tag{5.7}$$

with integer n. The width of the instability region of the first parametric resonance is given by

$$-\sqrt{\left(\frac{q\omega_z}{2}\right)^2 - 4\gamma^2} < \Delta < \sqrt{\left(\frac{q\omega_z}{2}\right)^2 - 4\gamma^2} \tag{5.8}$$

where Δ is the detuning from resonance. The time evolution of a single particle subject to parametric drive is in general not a simple sinusoidal oscillation, however the Floquet theorem states that the motion has the periodicity of the parametric drive. A general characteristic is the exponential growth of the amplitude of the oscillation over time within the instable regions, which can be expressed as

$$b(t) = \delta z_t e^{\left(\frac{1}{4}\sqrt{q^2\omega_z^2 - 4\Delta^2} - \gamma\right)t}. \tag{5.9}$$

Parametric oscillations have a threshold that increases for higher orders according to

$$q_n \approx \left(\frac{2\gamma}{\omega_z}\right)^{1/n}. \tag{5.10}$$

For a cloud of atoms in an isotropic trap, the lowest order parametric resonance corresponds to the radial breathing mode, where the whole cloud expands and contracts periodically. As was discussed in chapter 2.2.2, interactions shift the resonance frequency of this mode to higher frequency. Besides the resonances given by the Mathieu equation that exist on a single particle level, there are resonance series of shape oscillations for clouds of atoms (see chapter 2.2.2). For efficient excitation of a mode, the trap modulation has to match both the symmetry properties and the eigenfrequency of the mode. In our experiments, the magnetic traps have cylindrical geometry with the long axis aligned parallel to the surface. The modulation will thus match the symmetry of radial quadrupole modes.

5.1.3. Coupling Hamiltonian

The atomic c.o.m. mode represents an undamped mechanical oscillator that can be routinely prepared in the ground state. One goal is to achieve this also for micro- and nanomechanical resonators. In this regime, a coupled atom-resonator system has to be described quantum mechanically. Using the results of the previous sections, the Hamiltonian for the combined system can be approximated by

$$H = \frac{P^2}{2M_{\text{eff}}} + \frac{1}{2}M_{\text{eff}}\omega_m^2\delta z_c^2 + \sum_i \left(\frac{p_i^2}{2m} + \frac{1}{2}m\omega_z^2(z_i - \delta z_t)^2\right), \quad (5.11)$$

which represents the kinetic and potential energy of the resonator and the atoms, and where z_i is the position of the i-th atom and δz_c the excursion of the resonator from the static position z_c. We have chosen the trap minimum position as the origin and neglected the modulation of ω_z. The coupling shows up in the potential term of the atoms, which can be expanded to

$$H_{\text{pot}} = \sum_i \frac{1}{2}m\omega_z^2(z_i - \epsilon\delta z_c)^2 = \sum_i \frac{1}{2}m\omega_z^2(z_i^2 - 2\epsilon z_i \delta z_c + \epsilon^2 \delta z_c^2). \quad (5.12)$$

The first term on the right hand side is the unaffected atomic potential, the second term describes the coupling, and the third is a correction to the resonator frequency, which can be neglected.

To derive a quantized Hamiltonian, the amplitudes $z_i, \delta z_c$ have to be quantized by setting $\hat{z}_i = z_{\text{qm}}(\hat{b}_i^\dagger + \hat{b}_i)$ and $\delta \hat{z}_c = a_{\text{qm}}(\hat{a}^\dagger + \hat{a})$ with the ground state amplitudes $z_{\text{qm}} = \sqrt{\hbar/2m\omega_z}$, $a_{\text{qm}} = \sqrt{\hbar/2M_{\text{eff}}\omega_m}$ and the bosonic creation and annihilation operators $\hat{a}^\dagger, \hat{b}^\dagger, \hat{a}, \hat{b}$. This casts the coupling term of Equation 5.12 in the form

$$\hat{H}_{\text{int}} = \sum_i m\omega_z^2 \epsilon \hat{z}_i \delta \hat{z}_c = \sum_i \frac{1}{2}\epsilon\hbar\omega_z\sqrt{\frac{m}{M_{\text{eff}}}}(\hat{b}_i^\dagger + \hat{b}_i)(\hat{a}^\dagger + \hat{a}). \quad (5.13)$$

This is a linear coupling in the cantilever amplitude, which can be further simplified in the rotating wave approximation (neglecting the off resonant terms $\hat{a}^\dagger\hat{b}^\dagger$ and $\hat{a}\hat{b}$) to $\hat{H}_{\text{int}} = \hbar g_0 \sum (\hat{a}_i^\dagger \hat{b} + \hat{a}_i \hat{b}^\dagger)$. Here we have introduced the single phonon coupling strength g_0. For the center of mass mode $\hat{b} = 1/\sqrt{N}\sum_i \hat{b}_i$ one obtains a collectively enhanced coupling strength

$$g_N = \sqrt{N}g_0 = \frac{\epsilon\omega_z}{2}\sqrt{N}\sqrt{\frac{m}{M_{\text{eff}}}}. \quad (5.14)$$

The coupling is linear in ϵ and contains a disadvantageous term $\sqrt{m/M_{\text{eff}}}$. E.g. for the microcantilever used in our experiment, this term is of the order of 10^{-7}, and with typical parameters ($N = 1000, \epsilon = 0.13, \omega_z = 2\pi \times 10$ kHz) we obtain $g_N = 2\pi \times 1$ mHz, too small to be detected. However, for a thermally driven cantilever at room temperature, the coupling would lead to a phonon transfer rate of $2g_N\sqrt{n_{\text{th}}} \approx 2\pi \times 200$ Hz, where $n_{\text{th}} \cong k_B T/\hbar\omega_z$ is the thermal phonon occupation of the resonator. This suggests that thermal motion should lead to an observable coupling.

5.2. Measurement of atom loss in the surface potential

In a first set of measurements we determine the range of atom-cantilever distances $d = z_{t,0} - z_c$ where the atoms are affected by the surface potential U_s. We use a method similar to Lin et al. [24], where atom loss at the surface is used to obtain information about the trap depth U_0. Together with a simulation of the magnetic trapping potential U_m this allows us to extract information about U_s. In these measurements, the static properties of U_s are studied and the cantilever is undriven.

We prepare BECs of $N = 2.0 \times 10^3$ atoms with no discernible thermal component in a trap at $d = 16.6$ μm with $\omega_z/2\pi = 2.7$ kHz (see chapter 4.3 and 4.4.3 for trap parameters and trap characterization). At this d, we observe no influence of the surface. The trapping potential is then compressed to $\omega_z/2\pi = 10$ kHz (5 kHz), resulting in a BEC radius of 290 nm (430 nm), and ramped adiabatically within 1 ms (3 ms) to a set value of $z_{t,0}$ close to the cantilever surface.

To prove the adiabaticity of the ramp, we perform measurements for various ramp times and compare the shape and position of surface loss data. Additionally, we analyze residual cloud oscillations after ramp back into the detection trap. We measure oscillation amplitudes of $b' = 4 - 7$ μm after $dt = 4$ ms TOF from a detection trap with $\omega_{z,d}/2\pi = 750$ Hz. Assuming that the excitation originates only from the ramp to the surface, we obtain a maximum oscillation amplitude of $b_r = \frac{\omega_{z,d}}{\omega_z}\frac{b'}{\omega_{z,d}dt} = \frac{b'}{\omega_z dt} = 28$ nm in the trap at the cantilever. More details about the ramping can be found in Appendix B.

The atoms are held at the cantilever for an interaction time $t_h = 1$ ms during which some of the atoms are lost because of the reduced U_0. The short ramp and holding time is chosen in order to minimize the influence of surface evaporation (which is not very well quantified), technical heating, and three body collisional loss (see chapter 4.4.2). After the interaction, the atoms are ramped back into a relaxed trap at large distance where the remaining atom number N_r is determined by absorption imaging. Figure 5.3 (a) shows a typical absorption image. A measurement of the remaining atom number as a function of $z_{t,0}$ is shown in Fig. 5.3 (b). The atoms are lost in a small interval of a few hundred nm at a distance of ~ 66 μm from the chip surface.

5.2.1. Determination of the cantilever position

The expected position of the cantilever is roughly estimated to be $z_c = 68 \pm 10$ μm. This is obtained from the knowledge of the spacer chip thickness (48 ± 5 μm) on which the cantilever is glued, and the thickness of the glue layers. Here, z_c refers to the position of the surface of the metallized side of the cantilever. The surface of the dielectric side is located at $z_c - h$, where $h = 450 \pm 40$ nm is the cantilever thickness specified by the manufacturer and confirmed in electron microscope images.

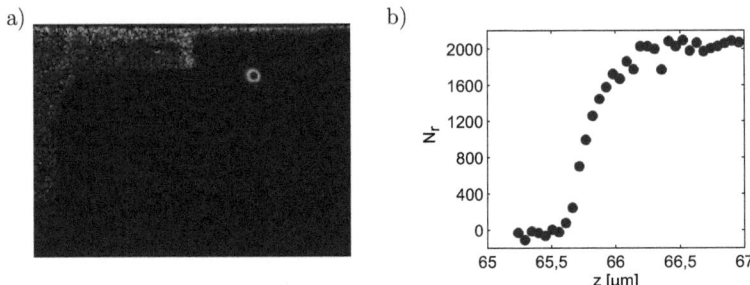

Figure 5.3.: (a) Absorption image of a BEC of 2000 atoms after 1.8 ms time of flight from the detection trap. The structure on the left is the shadow of the cantilever support chip with the cantilever located at the upper border of the image. (b) Remaining number N_r of atoms in a trap with $\omega_{z,0} = 10$ kHz after $t_h = 1$ ms at the surface.

For a more precise determination of z_c we use several measurements of atom loss in the surface potential as shown in Figure 5.3 (b). In these measurements, the position of the magnetic trap minimum, $z_{t,0}$, is obtained from a simulation of the magnetic trapping potential U_m (see chapter 4.4.3). We check the simulated U_m by comparison with measurements of the trap frequencies, the magnetic field at the trap bottom, and the trap position in absorption images. From this we estimate a relative uncertainty in $z_{t,0}$ of $\pm 3\%$. This leads to an absolute uncertainty of ± 2 µm at $z_{t,0} = 65$ µm, which is also the absolute uncertainty in the z-axis in Figure 5.3 (b).

Taking advantage of the suspended structure we can perform surface loss measurements also on the dielectric side of the cantilever, using the atoms as a "caliper" that measures the effective cantilever thickness including U_s. This involves manoeuvering around the cantilever with several magnetic field ramps to the point of measurement, and back again for imaging. We use a trajectory that orbits the cantilever at its width along the x-axis. It involves an axial displacement of the cloud of ~ 300 µm by currents in a neighboring dimple wire. It is advantageous to transport a pre-cooled thermal cloud around the cantilever and to prepare the condensate between cantilever and chip. After the measurements, the atoms have to be transported back around the cantilever. More details about the "manoeuver" can be found in Appendix C. We have no clear signature of the phase transition in the traps at the back side of the cantilever, in part due to heating during the transport back to the detection trap. Using surface loss data for thermometry (see section 5.3.1 below) we find that during the measurements at the cantilever the clouds have a temperature $\sim 50\%$ higher than for the measurements on the directly accessible metallized side.

5.2 Measurement of atom loss in the surface potential

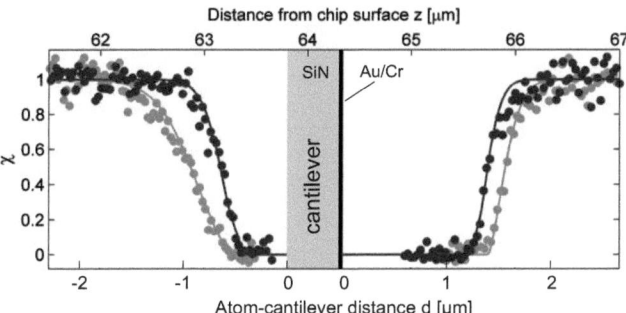

Figure 5.4.: Fraction χ of atoms remaining in the trap after $t_h = 1$ ms at distance d from the cantilever surface. Black (grey) data points correspond to a trap with $\omega_{z,0}/2\pi = 10.0$ kHz (5.1 kHz). Solid lines: Fit with a surface loss model, for details see chapter 5.3. The extracted cantilever position is shown.

Figure 5.4 shows the remaining fraction $\chi = N_r/N$ as a function of d for both sides of the cantilever and for two different trap frequencies. Because the cantilever has to lie somewhere in the region where $\chi = 0$, this allows us to determine the absolute cantilever position $z_c = 64.7 \pm 2.1$ μm with an error comparable to the uncertainty in $z_{t,0}$. However, we point out that the uncertainty in d is much smaller than the absolute uncertainties in $z_{t,0}$ and z_c. This is so because the distance between magnetic traps right above and below the cantilever is known to ± 60 nm (corresponding to the $\pm 3\%$ relative uncertainty in $z_{t,0}$). For perfectly known surface potentials U_s, this would also be the uncertainty in d.

Comparing the data with a simulation of the total potential U allows us to calibrate d to ± 160 nm and to obtain information about U_s, because U_m is very well known. We find that the distance between the two points where the atoms get lost is larger than expected from the cantilever thickness and the strength of the Casimir-Polder potential on both sides. From an analysis of surface loss and resonant coupling measurements we deduce that $U_s \approx U_{\text{CP}}$ on the dielectric side, while on the metallized side U_s is stronger by about two orders of magnitude, most likely due to surface adsorbates (for details on the analysis of the surface potential and the calibration of d see chapter 5.4).

We can model the loss of atoms in the surface potential U_s by a sudden truncation of the Boltzmann tail of the residual thermal cloud coexisting with the condensate, in combination with 1D evaporation and tunneling [24] (for details on the model see chapter 5.3). When comparing with the data, we leave the cloud temperature T as a free parameter. The data is fit best for $T = (1.5, 1.2, 1.0, 0.6) T_c$ (solid lines from left to right in Fig. 5.4), where the critical temperature T_c is calculated by Eq. 2.19.

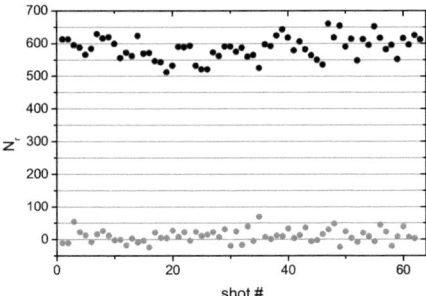

Figure 5.5.: Atom number noise on the slope of a surface loss curve. Black data: Repetitive measurement of the atom number for constant trap parameters $\omega_z/2\pi = 10.5$ kHz, $d = 1.3$ μm, and $t_h = 3$ ms. The mean atom number is $\langle N \rangle = 585$ and the overall noise amounts to $\sigma = 34$. Grey data: Apparent atom number, determined in an area of same size as used for the determination of N but without atoms present. This yields an imaging noise of $\sigma = 19$ atoms.

These values are systematically larger but still in reasonable agreement with those from independent measurements of T in time-of-flight for corresponding traps on the metallized side. We note that for the short t_h of the measurements in Fig. 5.4, the effect of evaporation and tunneling is small, and our determination of the cantilever position is independent of t_h.

5.2.2. Positioning reproducibility

To estimate how reproducible atoms can be positioned close to the surface, we use the fact that the trap position translates into atom number on the slope of a surface loss curve. The data shows that e.g. for a 10 kHz trap, atoms get lost over a distance of only 300 nm with a maximum slope of 4.5 atoms/nm for the analysed dataset. For repetitive measurements of the atom number after ramping to $d = 1.3$ μm where $\chi = 0.4$ and $\langle N \rangle = 586$ we find an rms noise $\sigma = 34$. Correcting for the noise of the imaging system $\sigma_I = 19$ that is observed when no atoms are present, we obtain $\sigma_N = \sqrt{34^2 - 19^2} = 28.2$, close to the value $\sqrt{586} = 24.2$ for shot noise.

As a worst case estimate we attribute all the noise to fluctuations of $z_{t,0}$, which yields $\Delta z_{t,0} = 6.3$ nm rms. We can compare this value to the positioning reproducibility expected from the stability of the magnetic potentials. The trap position is defined by a current through the Ioffe wire with a relative current stability of $\Delta I_I/I_I = 1.5 \times 10^{-5}$ and a homogeneous field along the y-axis, having $\Delta B_{b,y}/B_{b,y} = 5.2 \times 10^{-5}$. Together, the two sources define the trap position with a

5.3 Model for surface induced atom loss

rms stability $\Delta z_{t,0} = \mu_0/2\pi\sqrt{(\Delta I_I/B_{b,y})^2 + (I_I \Delta B_{b,y}/B_{b,y}^2)^2} = 3.3$ nm. A larger contribution comes from the background magnetic noise in the lab of $\Delta B = 20$ mG peak to peak, which in the worst case (when pointing exactly along our y-axis) would lead to $\Delta z_{t,0} = 7.4$ nm rms. Thus, the estimated positioning reproducibility is at the level expected from background magnetic field fluctuations. By implementing a magnetic shield and using a better current source for the $B_{b,y}$ field, position fluctuations could be brought down below 1 nm. However, at this level also the mechanical stability of the coils becomes relevant. E.g. if our $B_{b,y}$-coils drift in position by 1 μm along (x, y, z), the relative change in the field is $\Delta B_{b,y}/B_{b,y} = (3.1, 0.1, 9.3) \times 10^{-6}$ and the trap shifts by $\Delta z_{t,0} = (0.2, 0.06, 0.6)$ nm for our trap parameters. A temperature change of only 1°C gives rise to a differential position shift along z between the atom chip and the coils by ~ 2 μm due to the different thermal expansion of the Pyrex vacuum cell ($\alpha_{ex} = 3 \times 10^{-6}$ /K) and the steel coil mounts ($\alpha_{ex} = 13 \times 10^{-6}$ /K).

5.3. Model for surface induced atom loss

Here we discuss the model for the loss of atoms in the attractive surface potential U_s of the undriven cantilever. A simple model describing such measurements was developed by Lin et al. [24] and summarized in chapter 2.4.2. When the magnetic trap is ramped to the surface, the trap depth is reduced to U_0 by the surface potential. The model assumes that this leads to a sudden loss of atoms with energy $E > U_0$. For a thermal cloud this is described by the truncation of the tail of the Boltzmann distribution, while for partially condensed clouds at temperature $T > 0$, only the residual thermal cloud coexisting with the condensate is affected. Furthermore, it includes 1D evaporation from the trap to account for collisional repopulation of the high energy states. In summary, the remaining fraction of atoms in the trap is given by

$$\chi = (1 - e^{-\eta})e^{-\Gamma(\eta)t_h}, \tag{5.15}$$

where $\eta = U_0/k_B T$ is the ratio of the trap depth and the thermal energy, and $\Gamma(\eta)$ is the 1D evaporation rate according to equation 2.80. Evaporation is important when $t_h \gg \tau_{el}$. In the measurement of figure 5.4, $\tau_{el} = 0.2 - 0.6$ ms and $t_h = 1$ ms, and evaporation has only a small effect.

Finally, atoms can be lost from the trap by tunneling through the barrier. The loss rate can be estimated by calculating the transmission coefficient $T(E, U)$ of the barrier in the WKB approximation according to equation 2.82. This determines the tunneling rate $\Lambda = \omega_z/2\pi T(E, U)$ (Eq. 2.83). For our measurements, tunneling will contribute when $\Lambda(E) \approx 1/t_h$. As Λ increases exponentially with E, the effect can be accounted for by setting the potential depth to the value where the tunneling rate equals $1/t_h$, or $U_\text{eff} = E(\Lambda(E) = 1/t_h)$.

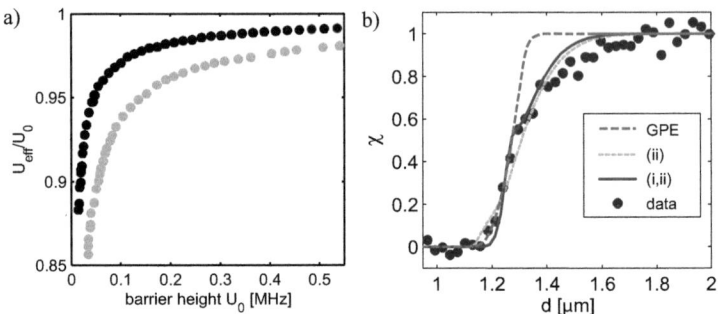

Figure 5.6.: (a) Calculated relative reduction of the barrier height U_{eff}/U_0 due to tunneling as a function of U_0 for $\Lambda(U_{\text{eff}}) = 1$ kHz. Black (grey) circles correspond to a trap with $\omega_z/2\pi = 10$ kHz (5 kHz). (b) Measured fraction of remaining atoms compared to a numerical simulation of the Gross-Pitaevskii equation, to the surface loss model with modification (ii) and $T = T_c$, or both (i) and (ii) with $T = 0.8\,T_c$. Parameters are $\omega_z/2\pi = 10$ kHz, $t_h = 1$ ms, $U_s = U_{\text{CP}} - 200C_4/(z-z_c)^4$.

Figure 5.6 (a) shows a numerical simulation of the relative trap depth reduction U_{eff}/U_0 as a function of the barrier height U_0 for $t_h = 1$ ms. The behaviour of the (absolute) barrier reduction $\Delta U = U_0 - U_{\text{eff}}$ can be approximated by a polynomial of the form $\Delta U \cong c U_0^{0.3}$, and for the traps used in the experiments we find $c(5\text{ kHz}) = 5.6 \times 10^{-3}$ and $c(10\text{ kHz}) = 1.3 \times 10^{-2}$. The effect is of the order of a few percent and becomes important only for very shallow traps ($U_0 < 100$ kHz). Overall, tunneling shifts surface loss curves by $z \sim 20-60$ nm compared to the case where tunneling is neglected. It thus has only a small impact for the interpretation of atom loss at the surface.

5.3.1. Improvements of the model

Although this simple model already describes our data fairly well, several improvements and alternative approaches are possible:

(i) The model describes loss only from the thermal cloud. A more accurate description can be obtained by assuming a bimodal cloud for $T < T_c$ and including also loss from the condensate. The condensate and thermal atom numbers are obtained by equation 2.20 or 2.21. The thermal cloud is lost for $U_0 \geq \mu_c$ as described above with a modified $\eta = (U_0 - \mu_c)/k_B T$. The condensate is lost for $U_0 < \mu_c$, where the number of remaining atoms N_r can be determined from $\mu_c[N_r] = U_0$, resulting in $\chi_{\text{BEC}} = (U_0/\mu_c)^{5/2}$.

(ii) For $T \sim T_c$, most of the surface loss occurs for $\eta \leq 1$, where the simple

5.3 Model for surface induced atom loss

evaporation law is no longer valid and leads to unphysically large Γ. We correct this by introducing a cutoff at the cross dimensional mixing rate [267], which we implement by setting

$$\Gamma^{-1} = \tau_{el}\left(\frac{1}{f(\eta)\exp(-\eta)} + 2.7\right). \tag{5.16}$$

However, this is only a simple patch and does not account for the qualitatively different situation at small η.

(iii) The repulsive interaction between the condensate and the thermal cloud pushes the latter out of the trap center (see equations 2.46 and figure 2.6). This leads to a broadening of the loss curves. This effect could be included by an effective potential

$$U_{\text{th}}(\mathbf{r}) = \left|\frac{1}{2}m(\omega_x^2 x^2 + \omega_y^2 y^2 + \omega_z^2 z^2) - \mu_c\right| \tag{5.17}$$

for the thermal cloud. However, this no longer permits to treat the sudden loss in the manner discussed above.

(iv) In chapter 5.8.2 we use a numerical 1D simulation of the Gross-Pitaevskii equation to study the dynamics of a BEC in a modulated trap. In the context here, we can use it to simulate the loss in the static surface potential.

(v) Technical heating, three-body collisional loss, and cooling due to evaporation of the atoms are not included in the model. These effects have a strong dependence on the trap frequency and in part also on the atom number.

Figure 5.6 (b) shows a comparison of two of the modifications of the model together with a numerical simulation for a pure BEC (iv) and experimental data. The measurements often display a characteristic kink, which indicates a partially condensed cloud with different loss behaviour of the thermal cloud and the condensate fraction. The data can be reproduced by using the improvements (i) and (ii).

When applying the above improvements (i) - (iv), we find that the resulting change in the calibration of the atom-cantilever distance d is ± 80 nm, which is within our error bar on d. Furthermore, we point out that the data can also be analyzed without a detailed model for atom loss by simply exploiting that $\chi = 0$ corresponds to the values of the atom-cantilever distance d where the trap has vanished (which is well described by the condition $U_0 < \hbar\omega_z/2$). This analysis depends only on the knowledge of the trapping potential U, and again yields similar results as the model described above for the short t_h of the measurements in Figure 5.4, where evaporation does not play an important role.

5.3.2. Heating rate analysis with the surface loss model

In chapters 2.5.3 and 4.4.2 we have discussed limitations due to technical heating in the trap. A difficulty is that temperature and heating rate measurements in TOF

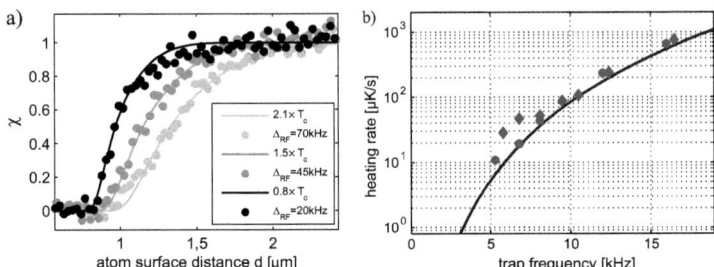

Figure 5.7.: (a) Surface loss measurements for different cloud temperatures. The temperature is set by changing the final value of the radio frequency for rf-evaporation. For the data we quote the detuning of the RF stop-frequency Δ_{RF} from the trap bottom $\nu_{RF,0} = 4.950$ MHz. The fitted temperatures are slightly higher but in reasonable agreement with the temperature following from the RF stop-frequency ($T = (2.0, 1.3, 0.6) \times T_c$). Such a reference measurement serves as "calibration" of the thermometer. (b) Heating rates extracted from the surface loss model applied to several measurements with given initial temperature and varying holding time. The solid line is the heating rate calculated with Eq. 2.99 for the measured current noise level.

show large uncertainty for small clouds. In an alternative approach we can use the fit results of the surface loss model to obtain the temperature of the cloud. Even though the accuracy of the evaluated temperature might be worse due to systematic errors of the model, the reproducibility of the determination is very good. When applied to measurements with varying interaction time t_h, the temperature as a function of time and thereby a heating rate can be extracted. In Figure 5.7 (a), surface loss measurements are shown where different cloud temperatures have been set by changing the final value of the radio frequency during rf-evaporation. The measurement illustrates, how well temperature differences can be discerned. From repetitive measurements with fixed parameters we find a variation of the resulting temperature $\Delta T = \pm 0.1\, T_c$. Note that the only free parameters of the fit are the position of the surface (which is the same for all fits in the figure) and the temperature.

To extract heating rates, we perform measurements for fixed initial temperature with varying holding time at the surface for several trap frequencies. The results are summarized in Figure 5.7 (b) together with a prediction of trap heating. The observed heating rate dependence follows a ω_z^4 scaling as expected from technical noise induced heating (Eq. 2.97, 2.99), and can be quantitatively explained by the current noise of the source that drives the magnetic coils for the $B_{b,y}$ offset field (FUG 15A 20V, see chapter 4.4.2).

5.4. Analysis of the surface potential

Here we describe how we use static and dynamic surface loss measurements to obtain information about the surface potential $U_s = U_{\rm CP} + U_{\rm ad}$ and a calibration of the atom-cantilever distance $d = z_{t,0} - z_c$ on both sides of the cantilever.

5.4.1. Additional potential $U_{\rm ad}$

If we assume for the moment that only the CP-potentials are present on both sides of the cantilever (i.e. $U_{\rm ad} = 0$ on both sides), we would expect from a simulation of $U = U_m + U_{\rm CP}$ that the "effective cantilever thickness" $h_{\rm eff}$, defined by the width of the window where $\chi = 0$ in Figure 5.4, is $h_{\rm eff} = 1.4$ μm for $\omega_z/2\pi = 10$ kHz and $t_h = 1$ ms. However, we observe $h_{\rm eff} = 2.2$ μm. This shows that U_s is significantly stronger than the expected contribution from $U_{\rm CP}$ on at least one side of the cantilever. We explain this by the presence of an additional potential $U_{\rm ad}$ due to surface inhomogeneities or contamination [268, 129, 25, 26, 27, 269]. Without taking into account further information about $U_{\rm ad}$, this leaves an uncertainty in d of ± 400 nm, corresponding to the difference between the observed and expected $h_{\rm eff}$.

The atoms could be used as a three-dimensional scanning probe that allows one to map out the spatial dependence of $U_{\rm ad}$ in detail and to determine whether it is due to magnetic, electrostatic, or other interactions, see e.g. the measurements in [25, 26, 27, 269]. As the characterization of $U_{\rm ad}$ is not the main focus of our work, we simply determine its strength relative to $U_{\rm CP}$ in the relevant range of d by combining the measurements from chapter 5.2 (see Figure 5.4) with information from measurements of resonant atom-cantilever coupling as described in chapter 5.5 (see e.g. Figures 5.8, 5.11).

Such dynamical coupling measurements are performed on both sides of the cantilever in traps with similar U_0. We can determine U_0 to 10% from the measured curves in Figure 5.4 without detailed knowledge of U_s or d. Comparing measurements on both sides of the cantilever, we find that the dynamical coupling signal is a factor $\beta = 3.2 \pm 0.6$ larger on the metallized side. Furthermore we observe a linear dependence of the coupling signal with the cantilever amplitude and thus also a linear dependence of the signal on $\delta z_t = \epsilon a$. From this we conclude that ϵ has to be larger by the same factor, and the ratio β is thus given by the ratio of the coupling strength parameters on the metallized and dielectric side, $\beta = \epsilon_{\rm met}/\epsilon_{\rm diel}$. Because $\epsilon \propto \partial_z^2 U_s$, this implies a stronger surface potential on the metallized side. Stronger U_s also implies larger d to maintain the same U_0. Due to the fast decay of U_s with d, a substantially larger U_s is required on the metallized side (not just larger by a factor of order β).

5.4.2. Iterative determination of the strength of U_{ad}

To obtain an absolute value for the strength of the surface potential on both sides, we use an iterative procedure to match both the observed effective cantilever thickness and the measured coupling strength ratio β. We first choose a certain C_{ad} and then evaluate ϵ and d for the given U_0 on the dielectric side. This fixes the cantilever position z_c. Then we adjust C_{ad} on the metallized side to be consistent with the surface loss curves in Figure 5.4 and extract ϵ for the given U_0 on this side. We compare the values of ϵ on both sides and start a new iteration with weaker (stronger) C_{ad} on the dielectric side if their ratio is smaller (larger) than the observed β, or finish if it equals the observed β.

The observed $\beta = 3.2$ as well as the surface loss curves can be best explained by potentials of the form $U_{\mathrm{ad}} = C_{\mathrm{ad}}/(z - z_c)^4$ with the following coefficients:

$$\begin{aligned} C_{\mathrm{ad}} &= (200 \pm 100)\, C_4 & &\text{metallized side} \\ C_{\mathrm{ad}} &= (10 \pm 10)\, C_{4,d} & &\text{dielectric side.} \end{aligned} \quad (5.18)$$

For these potentials, $z_c = 64.36$ μm results.

For the metallized side, we use the C_4 coefficient for a perfect conductor according to Eq. 2.57, which bears a small error (on the percent level) due to the small thickness and the finite conductivity of the Au/Cr film. On the dielectric side, the thin SiN layer together with the Au/Cr film acts as a cavity or waveguide for the vacuum modes [127], which results in a correction to the CP-potential (see chapter 2.3.1). With Eq. 2.63 we calculate that at $d = 1.0$ μm this leads to a 25% larger potential than that of a bulk dielectric described by $C_{4,d}$ (see Eq. 2.58). On this side, the inferred potential is thus consistent with a pure Casimir-Polder potential.

To check the robustness of our analysis against changes in the assumed distance dependence of U_{ad}, we perform similar analyses with other distance-dependences, such as $U_{\mathrm{ad}} \propto (z - z_c)^{-3}$ on both sides or $U_{\mathrm{ad}} \propto (z - z_c)^{-4}$ on the dielectric side and $U_{\mathrm{ad}} \propto (z - z_c)^{-3}$ on the metallized side. These analyses result in similar calibrations of d. The overall error in d is ± 160 nm, which contains the uncertainty in U_{ad}, $z_{t,0}$, U_0, β, as well as the uncertainty in the cantilever thickness, and a contribution due to residual oscillations of the atoms in the trap due to the ramping to the cantilever.

We observe that U_{ad} slowly changes over time by up to a factor of four on a time scale of weeks. The measurements used to determine U_{ad} described above were all performed on the same day. The change in U_{ad} during the course of these measurements is negligible. However, for the analysis of measurements from other days, the strength of the surface potential is known with less precision. But as the relative change in the coupling strength ϵ is much less than the relative change in the surface potential strength (by a factor $\beta\, C_4/C_{\mathrm{ad}} = 1.6 \times 10^{-2}$), this affects only the atom-surface distance and not the dynamical coupling.

5.4.3. Adsorbates

A likely explanation of the observed $U_{\rm ad}$ are ^{87}Rb adsorbates deposited during operation of the experiment. This effect was studied in detail in [25, 26, 27, 269]. The electric dipole moment of Rb on gold, $\mu_{el} \approx 1 \times 10^{-29}$ Cm, is about one order of magnitude stronger than on SiN [25, 269]. Furthermore, as most of the measurements are performed above the metallized surface, we estimate the adsorbed atom number on this side to be substantially larger than on the dielectric backside of the cantilever. Both effects would lead to a stronger $U_{\rm ad}$ on the metallized side.

As discussed in chapter 2.3.2, the expected stray field of the spatial adsorbate distribution can be well described by a power law $U_{\rm ad}(z) = C_{\rm ad}/(z-z_c)^4$ in the region of interest. We find that $U_{\rm ad}$ matches the observed U_s for 3500 BECs containing 2000 atoms each, distributed over an area of 10×1 μm^2, about two times the size of a condensate. This is a realistic atom number, consistent with the number of experiments performed. The observed changes in the surface potential are consistent with the picture that atoms are deposited on the surface and subsequently diffuse or desorb again [269].

However, in chapter 2.3.2 we have pointed out that for our vacuum conditions with rather high background pressure of $p = 3 - 6 \times 10^{-10}$ mbar, the cantilever surface is expected to have a large Rb coverage due to adsorption from the Rb background gas. On the other hand, the heating of the cantilever by wire currents and the readout laser (see chapter 4.1.2) can lead to temperatures up to 40 °C and thus increase the desorption rate of Rb by almost one order of magnitude. The state of the surface is thus not very well known and the given estimate bears some uncertainty.

5.5. Detection of mechanical motion with BECs

We now describe our main experiments, where cantilever oscillations are coupled to the motion of trapped atoms nearby. We study mainly the c.o.m. mode of the atoms because it has the best mechanical properties and is relevant for future scenarios. For comparison, we also investigate the behaviour of the breathing mode as a representative for higher order collective excitations.

In our measurements we use trap loss as indicator for the coupling. This implies that the excitation has to lead to large amplitude collective oscillations, such that a part of the cloud spills over the barrier and is lost from the trap. Therefore the atoms are ramped to a distance, where the magnetic trap is markedly affected by the static surface potential. For typical distances used in the measurements, the coupling strength parameter ϵ reaches values of $\epsilon = 0.05 - 0.15$ and the trap depth is of the order $U_0 = 8 - 25 \times \hbar\omega_z$.

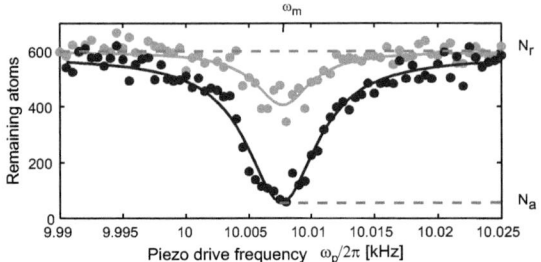

Figure 5.8.: Remaining atom number after $t_h = 3$ ms in a trap with $\omega_z/2\pi = 10.5$ kHz at $d = 1.5$ μm from the driven cantilever, for varying drive frequency ω_p. The dark (light) grey circles correspond to a cantilever amplitude $a = 120$ nm (50 nm) on resonance. Solid lines: Lorentzian fits with 6 Hz FWHM, corresponding to the width of the cantilever resonance. The remaining atom number with (N_a) and without (N_r) resonant piezo excitation of the cantilever is indicated.

5.5.1. Probing the cantilever fundamental mode spectrum

For a first signature of dynamical coupling between a BEC and the resonator we use the atoms to reveal the mechanical resonance spectrum of the fundamental mode of the cantilever. Therefore we excite the cantilever with the piezo at frequency ω_p. When ω_p is resonant with the cantilever's fundamental out-of-plane mode at $\omega_m = 2\pi \times 10$ kHz, the cantilever oscillates with an amplitude a of typically several tens of nm[1]. We prepare BECs on the metallized side at $d = 1.5$ μm in a trap with $\omega_z/2\pi = 10.5$ kHz, so that resonance $\omega_z \approx \omega_m$ is given, and let the atoms interact with the vibrating cantilever for $t_h = 3$ ms. In this trap, $U_0 = h \cdot 205$ kHz $= 9 \, \mu_c$ and $\chi = 0.4$ if the cantilever is undriven.

When ω_p is scanned from shot to shot of the experiment, a sharp resonance in the remaining atom number is observed for $\omega_p = \omega_m$, see Fig. 5.8. The width of the atomic resonance matches the width of the cantilever resonance very well, and it is thus possible to resolve the spectral response of the cantilever with the atoms.

Note that a is more than one order of magnitude smaller than d, and the cantilever does not touch the atomic cloud. A calculation of the potential yields that the surface potential of the oscillating cantilever modulates z_t with an amplitude $\delta z_t = 10$ nm (4 nm) for $a = 120$ nm (50 nm) on resonance, and thus $\epsilon = 0.08$. This excites coherent motion of the atomic center of mass (c.o.m.). For large c.o.m. amplitudes, the anharmonicity of the deformed trap and the reduced U_0 convert this motion into heating and loss.

---
[1] Due to temperature dependent frequency shifts of the cantilever, excitation is only resonant for a short time during the experimental sequence. For more details see chapter 4.1.2

5.5 Detection of mechanical motion with BECs

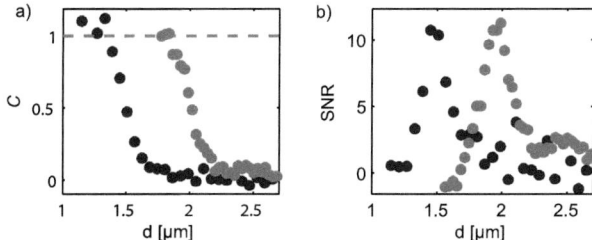

Figure 5.9.: (a) Contrast C and (b) signal to noise ratio SNR of the observed atomic signal as a function of atom-surface distance d. For this measurement we use constant $a = 90$ nm and $\omega_p = \omega_m$. Dark (bright) points correspond to $\omega_z/2\pi = 10.5$ kHz (5.0 kHz) and $t_h = 3$ ms (20 ms).

5.5.2. Distance dependence

An central property is the distance range over which the coupling is observable. To study this we measure N_r and N_a (see Fig. 5.8 for the definition) for various distances d. We perform the measurements for traps tuned to both the c.o.m and the breathing mode. For the latter mode, the resonance condition $\omega_m = 2\omega_z$ has to be fulfilled such that $\omega_z = 2\pi \times 5$ kHz in this measurement (see also chapter 5.7). We set the coupling time t_h to the respective optimal values (see chapter 5.5.3).

Figure 5.9 (a) shows the dependence of the atomic signal on d for a constant cantilever amplitude $a = 90$ nm. We show the contrast $C = (N_r - N_a)/N_r$ which measures the relative amount of coupling induced loss. To determine the signal visibility we calculate the signal to noise ratio SNR $= (N_r - N_a)/\sigma$, see Fig. 5.9 (b), with $\sigma = 34$ the r.m.s. atom number noise observed without cantilever driving. The strong variation of the signal over ~ 300 nm matches with the range of d where U_s modifies the trapping potential noticeably. This proves that no external mechanism such as e.g. direct driving of the atoms by the piezo, which is 4 mm away, contributes to the signal.

The sharp maximum of the SNR reveals a high sensitivity of the coupling on d, which thus requires high control over the atomic position. We find that the position and shape of the maximum depend on the coupling time and the exact trap frequency. To maximize the signal over the entire distance range, the trap frequency would have to be optimized for every set value of d, while here we set $\omega_{z,0}$ to a constant value over the measurement.

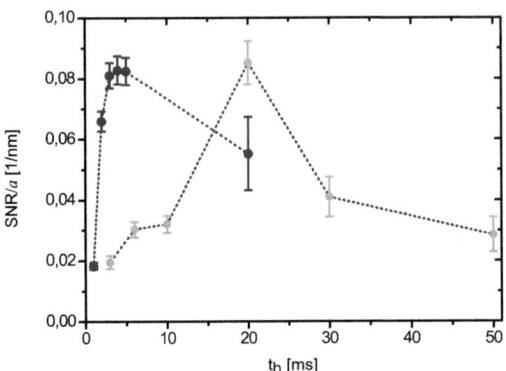

Figure 5.10.: SNR normalized to the cantilever amplitude a as a function of the hold time t_h at the cantilever. We compare the c.o.m. mode measured in a trap with $\omega_z/2\pi = 10$ kHz (dark), and the breathing mode with $\omega_z/2\pi = 5$ kHz (bright). For both modes the signal reaches a comparable maximal value of $\text{SNR}/a = 0.08$ nm^{-1}. However, the optimal t_h differs by a factor 5.

5.5.3. Dependence on hold time

The observed coupling signal depends non-trivially on the duration of the coupling t_h. For a cloud excited at the trap frequency in a perfectly harmonic trap, the linear increase in amplitude according to Eq. 5.3 suggests that t_h should be chosen as long as possible. Excitation of the breathing mode with exponential rise according to Eq. 5.9 calls for even longer interaction time. However, parasitic loss due to inelastic collisions, evaporation, or technical heating decrease N_r and thereby set an upper limit on t_h. Since part of this loss is density dependent, the optimum t_h will also depend on ω_z. A further influence comes from the anharmonicity of the trap, which reduces the amplitude growth due to dephasing and makes long coupling times unfavourable.

Figure 5.10 shows a measurement of the dependence of the SNR on t_h for both the c.o.m. and the breathing mode. We perform measurements at several distances and cantilever amplitudes and give the maximal value of SNR/a for each t_h. The error bars include the uncertainty in the determination of N_r and N_a as well as the uncertainty in a. The data shows that the c.o.m. mode is excited to large amplitude within a few ms, leading to a fast rise of the signal. We find an optimal interaction time $t_h = 4$ ms and a decrease of the signal for $t_h > 4$ ms with a time constant comparable to the lifetime of the atoms in the trap. Excitation of the breathing mode takes longer to lead to sizable atom loss. Yet, with the lower collisional loss and heating rates in the trap tuned to this mode frequency, the maximal SNR is comparable to that of the c.o.m. mode, but for longer hold time $t_h = 20$ ms. The comparable value of the optimum coupling signal for the two studied modes is rather a coincidence. It is a consequence of the trap frequency scaling of the parasitic loss, and thus depends on the trap geometry, the atom number N_r, the technical current noise level, and also on the value of the cantilever eigenfrequency.

5.6. Readout sensitivity

So far we have shown that resonant excitation of atomic motion via surface forces can lead to significant trap loss after a few ms of interaction time. This demonstrates that the atoms can be used for readout of cantilever motion. To determine the achievable sensitivity of the readout method we use parameters that maximize the SNR and measure the signal as a function of the cantilever amplitude.

Figure 5.11 shows measurements of the contrast as a function of the cantilever amplitude for the c.o.m. mode. For comparison, the signal for an off resonant trap with $\omega_z = 4$ kHz is shown. The measurements yield a minimum resolvable r.m.s. cantilever amplitude of $a_{\text{rms}} = 13 \pm 4$ nm for SNR=1 without averaging, where the error is dominated by the uncertainty of the cantilever amplitude (see chapter 4.1.2).

As introduced already in the previous chapter, we identify two origins for the observed sensitivity limit. First, the lifetime of the atoms in the trap set by parasitic loss limits the coupling duration and thus the achievable excitation for small amplitudes. This is indicated by the measurements shown in Fig. 5.10. Second, trap anharmonicity leads to dephasing of the cloud oscillations and thereby to a maximum cloud amplitude for a given cantilever amplitude. A detailed picture of the dynamics of the cloud and the influence of the trap anharmonicity is found by numerical simulations which we discuss in chapter 5.8.

We also perform coupling measurements on the dielectric back side of the cantilever in a trap with comparable trap frequency and depth U_0. We observe an approximately linear dependence $C \propto a$ on both sides as long as the contrast does not saturate, i.e. for $C < 1$. The coupling signal can thus be quantified by the value C/a, and we find it to be a factor $\beta = 3.2 \pm 0.6$ smaller on the back side than on the metallized side. Since the origin of the excitation is the modulation of δz_t, one can conclude that $C \propto \delta z_t$. With Eq. 5.1, the contrast is thus determined by the coupling strength parameter, $C/a \propto \epsilon = (m\omega_z^2)^{-1}\partial^2 U_s/\partial z^2$, and thereby related to the curvature of the surface potential. In chapter 5.4 we use this result together with static measurements to quantitatively infer the absolute strength of U_s on both sides of the cantilever.

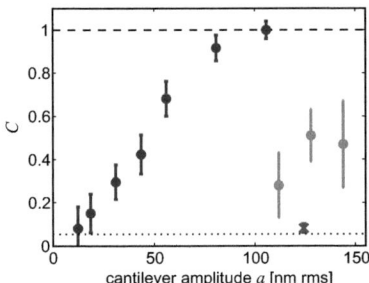

Figure 5.11.: Contrast C for the c.o.m. mode as a function of the cantilever amplitude a. The smallest detectable cantilever amplitude is $a_{\text{rms}} = 13 \pm 4$ nm for SNR=1 without averaging. For measurements on the metallized side (dark) we find $C/a = 1.1 \times 10^{-2}\text{nm}^{-1}$ while on the dielectric backside (bright), $C/a = 3.5 \times 10^{-3}\text{nm}^{-1}$, a factor $\beta = 3.2 \pm 0.6$ smaller. For comparison, the contrast for an off-resonant trap with $\omega_z/2\pi = 4$ kHz is shown (cross). The dotted line indicates the rms noise of the measurement.

Improvement of the sensitivity

We have used trap loss as the simplest way to detect BEC dynamics induced by the coupling. For trap loss to occur, the cantilever has to drive the BEC to large amplitude oscillations with $\sim 10^3$ phonons. Achieving such cloud amplitudes with small cantilever amplitudes is hindered by the strong trap anharmonicity close to the barrier, and by the finite trap lifetime. By contrast, BEC amplitudes down to the single phonon level could be observed by direct imaging of the motion. A coherent state $|\alpha\rangle$ of the c.o.m. mode of $N = 100$ atoms with $\alpha = 1$ released from a relaxed detection trap with $\omega_z = 2\pi \times 100$ Hz has an amplitude of $\sqrt{2\hbar\omega_z/mN}\alpha t = 400$ nm after $t = 4$ ms time-of-flight. This is about 10% of the BEC radius and could be resolved by absorption imaging with improved spatial resolution. From a simulation of the cloud excitation (see chapter 5.8) we estimate that $a_{\text{rms}} = 0.2$ nm would excite the BEC to $\alpha = 1$ within $t_h = 20$ ms and could thus be detected. This would be sufficient to resolve the thermal motion of our cantilever, which has a relatively large effective mass $M_{\text{eff}} = 5 \times 10^{-12}$ kg and correspondingly small r.m.s. thermal amplitude $a_{\text{th}} = \sqrt{k_B T/M_{\text{eff}}\omega_m^2} = 0.4$ nm, where $T = 300$ K is the cantilever temperature. Using similar cantilevers with comparable ω_m but smaller M_{eff} [199, 52], the thermal motion would be detectable already with the presently used technique.

Furthermore, one can harness the result that a stronger surface potential leads to a stronger coupling. Stronger potentials could be generated e.g. by electrostatic

5.7 Mode spectroscopy

charging of the cantilever. This should allow to reach ultimate values of $\epsilon \sim 0.3$ (see Fig. 5.2), thereby increasing the coupling strength by a factor ~ 4.

So far, the demonstrated and also the expected readout sensitivity for the type of cantilever as used here is not comparable to the achievements of optical readout techniques (see chapter 4.1.2 and 3.2). However, there are scenarios where the employed coupling mechanism opens new possibilities for readout or even manipulation of mechanical oscillators. One example is the application to ultra-light, nanoscale oscillators, which are difficult to access optically. We discuss this scenario in the outlook (chapter 6.1).

5.7. Mode spectroscopy

As discussed in section 2.2.2, the spectrum of collective mechanical modes is influenced by atomic collisions [101, 103] and shows additional, shifted resonance frequencies compared to the harmonic spectrum of a non-interacting gas. When anharmonicity is present, resonance broadening and nonlinear mode mixing can further modify the spectrum.

Studying the spectrum can give insight about which modes can be excited efficiently, how selective the atomic response is on the trap frequency, and to which extent the strong anharmonicity of the coupling trap affects the situation. We measure the dependence of the atomic response on the trap frequency ω_z and by this perform a spectroscopy of the strongest cloud excitations. We find that the cantilever can be coupled selectively to different, spectrally well separated BEC modes.

Experimental issues

In the ideal setting for such a measurement, the trap frequency is varied without affecting the number of bound levels of the trap $\sim U_0/\hbar\omega_z$ (or alternatively U_0/μ_c) nor changing the coupling strength parameter ϵ. Assuming that the temperature is proportional to ω_z and neglecting the dependence of the loss in the static surface potential on ω_z, this would result in a constant reference atom number $N(\omega_z) = N_0$ without driving. Then, the relative strength of resonances would be directly comparable and could indicate the relative strength of the modes. However, for the large range of trap frequency to be spanned, the change in the technical heating rate and the elastic and inelastic collision rate leads to vastly different trap lifetime (see chapter 4.4.2). A compromise is thus chosen to set the atom-surface distance $d(\omega_z)$ such that for a given cantilever amplitude all resonances remain unsaturated ($\chi > 0$) and that the changes in N_r are minimal (see lower panel of Fig. 5.12).

The holding time is chosen such that the observed resonance widths are not Fourier limited and that the SNR is close to the optimum as found in the measurements of chapter 5.5.3.

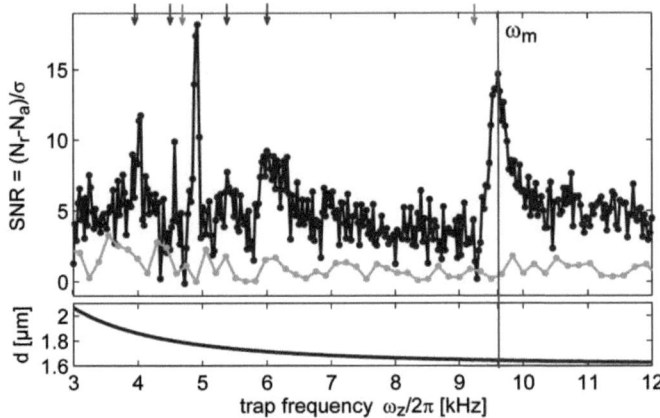

Figure 5.12.: Top graph: BEC response as a function of ω_z, for fixed $a = 180$ nm and $\omega_p = \omega_m$ (dark grey). Datapoints are connected with a line to guide the eye. Light grey: reference measurement without piezo excitation. We observe two major resonances at $\omega_m = \omega_z$ and $\omega_m = 2\omega_z$ and up to four smaller ones (dark arrows). Furthermore, we find reproducible anti-resonances with strongly suppressed signal (bright arrows). Due to cantilever aging, $\omega_m/2\pi = 9.68$ kHz in this measurement. Bottom graph: set values of d, chosen such that $N_r \approx$ const. ($N_r[10\text{ kHz}] = 700$, $N_r[5\text{ kHz}] = 1100$) and N_a does not saturate.

Loss spectrum

The cantilever is excited resonantly ($\omega_p = \omega_m$) to constant amplitude $a = 180$ nm with the piezo, and coupled to the BEC for $t_h = 20$ ms on the metallized side. In Fig. 5.12 we show how the observed atomic SNR changes when we scan ω_z.

The measured spectrum shows strong resonances at $\omega_m = \omega_z$ and $\omega_m = 2\omega_z$. They correspond, respectively, to the atomic c.o.m. mode and the high frequency $m_l = 0$ collective mode of the BEC in our cigar-shaped trap [101, 103]. These two modes have been studied in the previous chapters in detail. In our trap, the latter coincides also with the breathing mode of the thermal component of the gas. The mode at ω_z ($2\omega_z$) is excited by the cantilever through modulation of z_t (ω_z) and we calculate a modulation amplitude of $\delta z_t = 7$ nm ($\delta\omega_z = 2\pi \times 150$ Hz). The c.o.m. mode resonance at $\omega_m = \omega_z$ shows pronounced asymmetry and has a full width of 300 Hz. For the resonance at $\omega_m = 2\omega_z$, we observe a linewidth of only 60 Hz, corresponding to a quality factor of ≈ 100. This is close to the expected resonance width of both the instable region of a parametric resonance $\Delta\omega_z < \delta\omega_z/2 = 75$ Hz and the Fourier limit $\Delta\omega = 1/t_h = 50$ Hz. Furthermore, we find up to four weaker

5.7 Mode spectroscopy

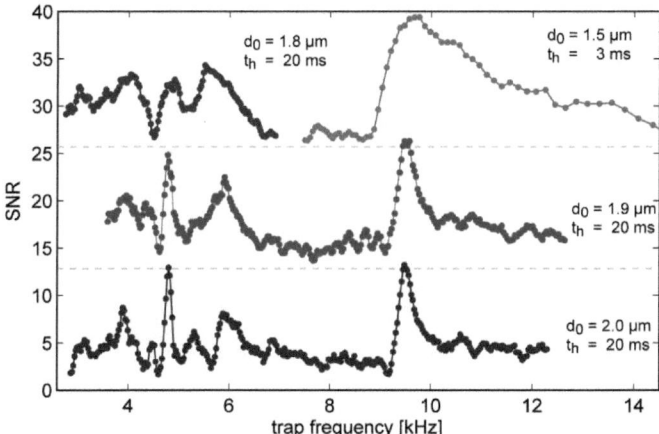

Figure 5.13.: Gallery of spectra taken at different distances and hold times. Spectra are shifted by a constant value of 13 (26) for better visibility and averaging of adjacent points is performed to reduce noise. To compare the chosen atom-surface distances of the measurements we quote the value $d_0 = d(4\text{ kHz})$, except for the measurement with $t_h = 3$ ms where $d_0 = d(10\text{ kHz})$. We observe resonance broadening for decreasing d and the relative height of the resonances changes. Notably, the BEC quadrupole mode at 6.1 kHz becomes stronger than the breathing mode at small d.

resonances at frequencies $\omega_m = (1.6, 1.8, 2.1, 2.4)\,\omega_z$. The first resonance can be identified with the $|m_l| = 2$ quadrupole mode of the BEC [101], whose frequency is given by $\omega_m = \omega_z\sqrt{2(1 + E_{\text{kin},\perp}/E_{\text{pot},\perp})}$ (see [103] and Eq. 2.43). We calculate the BEC kinetic energy $E_{\text{kin},\perp}$ and potential energy $E_{\text{pot},\perp}$ in the radial direction as in [100] for 1100 atoms, which reproduces the measured mode frequency.

Next to the resonances, we observe reproducible "anti-resonances" where the atomic response is suppressed by a factor of ~ 20. This can be used to switch the coupling on and off very efficiently by a slight detuning of the trap frequency. A possible origin for the suppression of atom loss is destructive interference between two resonances. This can arise e.g. due to a second resonance with nearby resonance frequency, either one of the indicated weaker resonances in the spectrum, or an oscillation along the transverse radial axis (the y-axis) in the trap. The two oscillations can lead to a beating and thus to destructive interference. Alternatively, the fast ramping to the surface can give rise to residual oscillations of the cloud. These can interfere destructively with the excitation by the cantilever close to an atomic resonance and thereby cause the anti-resonances. A similar effect has been observed e.g. in [213].

Distance dependence

We perform analogous measurements at smaller distances to see the effect of anharmonicity. Figure 5.13 shows examples of a few measurements. We observe broadening of the resonances, and the resonance at $\omega_m = 1.6\,\omega_z$ becomes stronger than the resonance at $\omega_m = 2\,\omega_z$ for small distance. We explain this by a complete removal of the thermal cloud for the small remaining barrier, which leaves only collective BEC modes. The $|m_l| = 2$ quadrupole mode matches the excitation symmetry best and becomes the strongest mode after the dipole mode. In a further measurement we study the behaviour of the c.o.m. mode for short holding time and significantly smaller d. We observe strong broadening and response for frequencies $\omega_z > \omega_p$ which corresponds to subharmonic excitation. In the 1D quantum simulation in chapter 5.8.2 we find such resonances, while they are absent in the classical simulation.

5.8. Simulation of cloud excitation

To confirm the interpretation of the measurements in the previous sections and to gain more insight into the dynamics of the atoms, we perform numerical simulations.

The surface potential introduces strong anharmonicity to the trap and leads to nonlinear response which is not predictable analytically. A proper description would have to deal with a solution of the generalized Gross-Pitaevskii equation in 3D e.g. in the Popov approximation. As this would be a major computational task, we restrict ourselves to two simple approaches.

In the first section, we simulate classical particles in the modulated trapping potential to approximate the behaviour of the thermal cloud. The evolution of individual trajectories show the effects of dephasing, the bounding of the atomic amplitude, and the frequency shift and broadening of the oscillation for large amplitude.

In the second section, we perform a simulation of the Gross-Pitaevskii dynamics of the condensate where we also explicitly include loss due to adsorption on the surface. We can thus model static and dynamic loss and study the distortion of the wave function, which is the origin of heating.

5.8.1. Simulation of classical trajectories

In this section we numerically solve the differential equation for a collection of point particles in a harmonic potential subject to a force arising from a surface potential of the form $U_s = -C/(z - z_c)^4$. The differential equation is given by

$$\sum_i \left(\ddot{z}_i + 2\gamma \dot{z}_i + \omega_0^2 z_i - \frac{4C}{m(z_i - z_c)^5} \right) = 0, \tag{5.19}$$

where z_i is the position coordinate of the i-th atom. The minimum of the magnetic potential is chosen to be at the origin $z_{t,0} = 0$, the cantilever position is $z_c =$

5.8 Simulation of cloud excitation

$z_{c,0} + a\cos\omega t$, and γ is the atomic amplitude damping rate which we neglect in the following. The temperature of the cloud enters via the choice of initial parameters $z_i(0), \dot{z}_i(0)$. For the calculation we use the ode45 solver of MATLAB.

As long as anharmonicities are negligible, we expect a linear rise of the oscillation amplitude on resonance as derived in chapter 5.1. We can use Eq. 5.3 for a crude estimate of the expected cloud amplitude after a coupling interval for typical experimental parameters. In our experiment, the minimum resolvable amplitude is $a = 13$ nm rms, measured in a trap with $\omega_z = \omega_m = 2\pi \times 10$ kHz at $d = 1.5$ μm. The analysis of the surface potential yields $U_s = 200 C_4/(z - z_c)^4$ (which is assumed from now on), and for the chosen distance results in $\epsilon = 0.13$ and thus $\delta z_t = 1.7$ nm for the minimum resolvable amplitude. After a coupling interval of $t_h = 20$ ms we expect an amplitude $b(20\text{ms}) = 200\pi\delta z_t = 1130$ nm. The trap has a radius (i.e. the distance between the trap minimum and the barrier) of ~ 700 nm, while the cloud has a $1/e$ radius $r_{th} = 150$ nm at $T = T_c/2$. A naive guess would be that loss becomes visible for an amplitude $b \sim 550$ nm. This indicates that deviations from Eq. 5.3 will be substantial and that the anharmonicity will dominate the dynamics.

Figure 5.14 shows the time evolution of a cloud at $T = T_c/2$ with a radius $r_{th} = 150$ nm in a resonant trap for typical coupling parameters as used in the experiments. For clarity we show a representative set of three classes of atomic initial conditions in the trap. The first class covers atoms with energy $E = p_i^2/2m + (1/2)m\omega_z^2 z_i^2 = k_B T_c/2$ with evenly spaced (p_i, z_i) to fill a ring in phase space. The second class represents atoms with $E = k_B T_c/4$ while the third class shows an atom initially at the trap bottom with $E = 0$. The simulation yields that a linear rise of cloud oscillation according to Eq. 5.3 is given only during the first ~ 2 ms. This is also the timescale during which all trajectories stay in phase, and for which the excitation can be considered as coherent or reversible. For longer times, the rim of the cloud begins to lag behind the cloud center, given by the lower curvature and hence longer oscillation period for larger amplitudes. For $t > 10$ ms, the oscillations of the individual trajectories have dephased completely and the excitation of the cloud corresponds to an increased temperature rather than a coherent displacement.

An important characteristic is the existence of a maximum amplitude $b_{\max}$ for a given excitation amplitude, a well known property of anharmonic oscillators [265]. Figure 5.15 (a) shows $b_{\max}(a)$ for an atom initially at the center of the trap. Only for amplitudes $a > 50$ nm the excitation suffices to kick atoms out of the trap. The upper bound is a consequence of the strong deformation of the trap close to the barrier. In our experiments, we observe coupling induced atom loss also for smaller amplitude. This is explained by the fact that the actual cloud extends up to the barrier and that atoms with energy close to U_0 are lost either directly due to the modulation of the barrier or excited within a few periods above the barrier.

Solving for $b_{\max}$ as a function of the trap frequency can be used to determine the spectrum of 1D excitations. Figure 5.15 (b) shows spectra for two different cantilever amplitudes with similar atom-surface distance as in the experiments. The

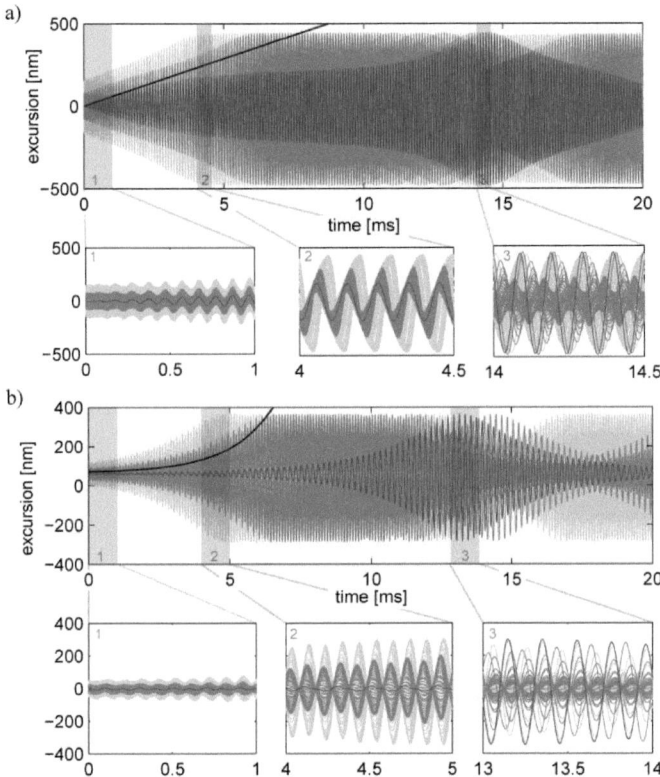

Figure 5.14.: (a) Resonant excitation ($\omega_m = \omega_z$): Atomic oscillation vs excitation time for $a = 13$ nm rms, $d = 1.5$ μm, and $U_s = 200 U_{\rm CP}$, resulting in $\epsilon = 0.13$ and $\delta z_t = 1.7$ nm rms. Bright grey lines: trajectories of atoms with $E = k_B T_c/2$; grey lines: atoms with $E = k_B T_c/4$; dark grey line: atom with $E = 0$. Dephasing due to the trap anharmonicity gives rise to a maximum amplitude $b_{\max} = 460$ nm of the atoms. The black line shows the linear rise in amplitude according to Eq. 5.3. Small panels show zooms into the shaded areas. (b) Parametric excitation ($\omega_m = 2\omega_z$) for $d = 1.7$ μm and $a = 70$nm, resulting in $\delta\omega_z \equiv \omega_z q = 408$ Hz. The black line shows the exponential growth given by Eq. 5.9.

5.8 Simulation of cloud excitation

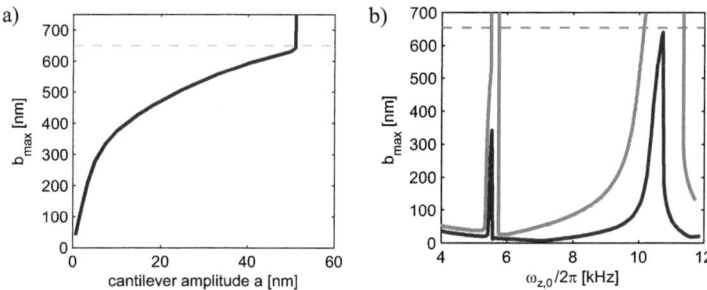

Figure 5.15.: (a) Maximum amplitude b_{max} as a function of cantilever amplitude a. For $b_{max} > 670$ nm the atom is lost in the surface potential. (b) Maximum amplitude as a function of trap frequency, analogous to the measurements of chapter 5.7. Parameters are $t_h = 20$ ms, $a = 50$ nm (dark), $a = 200$ nm (bright), and varying d between $1.5 - 1.9$ μm.

c.o.m. mode shows an asymmetry opposite to the experimental observation and has a width of ~ 300 Hz. As expected, it has a shape similar to a Duffing oscillator with softening anharmonicity.

We now study, to which extent coherent and reversible excitation of atomic motion is possible in the presence of anharmonicity. The condition that has to be met is that all trajectories remain in phase to a certain degree during a coupling interval. The degree of phase coherence then determines the fidelity of the c.o.m. excitation.

Dephasing is directly related to the dependence of the oscillation frequency on the oscillation amplitude. Figure 5.16 a shows a calculation of the frequency shift $\omega_{com}(b) - \omega_z$ as a function of oscillation amplitude for a trap with $\omega_{z,0}/2\pi = 10$ kHz at $d = 1.5$ μm according to equation 2.77. To test the validity of Eq. 2.77 for large amplitudes, we evaluate the oscillation frequency of trajectories obtained by the numerical solution of Eq. 5.2 by Fourier transform. We find that the spectrum becomes substantially broadened for large amplitude, owing to non-uniform motion. The analytical prediction is very accurate for small amplitudes, but underestimates the shift for amplitudes close to the trap radius.

For an extended cloud of atoms, individual trajectories will differ in amplitude and will thus oscillate at a frequency deviating from the frequency for the c.o.m. coordinate of the cloud. The consequence is a phase spread of the trajectories. For a quantitative estimate we consider the difference of the accumulated phase of two trajectories with minimum and maximum amplitude b_{min}, b_{max},

$$\Delta\phi(b,t) = \int_0^t \omega_{com}(b_{max})dt' - \int_0^t \omega_{com}(b_{min})dt', \tag{5.20}$$

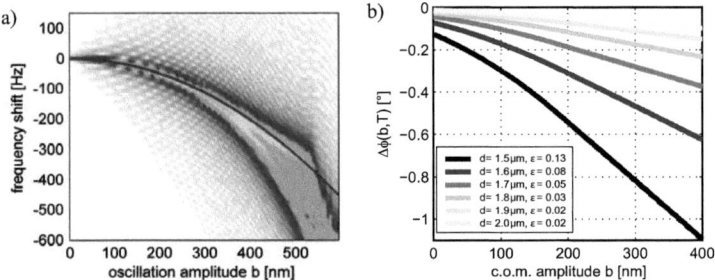

Figure 5.16.: (a) Oscillation frequency shift $\omega_{\text{com}}(b) - \omega_z$ as a function of oscillation amplitude calculated with Eq. 2.77 (solid line). The color scale shows a Fourier spectrum calculated from trajectories simliar to those in Fig. 5.14 for a trap with $\omega_{z,0}/2\pi = 10$ kHz, $d = 1.5$ µm, $t_h = 100$ ms, and $U_s = 200 \times U_{\text{CP}}$. (b) Phase spread increase per cycle $\Delta\phi(b, \mathcal{T})$ for various d and constant amplitudes $b_{\max} = b + r_{th}$ and $b_{\min} = \max(0, b - r_{th})$ and $r_{th} = 150$ nm.

where the two amplitudes are $b_{\min}^{\max} = \max(0, b \pm r_{th})$.

Figure 5.16 (b) shows the increase in phase spread per period as a function of the oscillation amplitude. For a typical coupling trap at $d = 1.7$ µm and a c.o.m. amplitude of the order of the cloud radius $b = 150$ nm, the per cycle phase spread amounts to $\Delta\phi = 0.15°$, such that after a holding time $t_h = 20$ ms or equivalently 200 oscillations, a phase spread of 30° has accumulated and the oscillation is already notably dephased. Note that due to its energy spread, a cloud at finite temperature yields a finite phase spread also for infinitesimal c.o.m. amplitude. Thus, also small amplitude oscillations show dephasing, and e.g. a collective single phonon excitation with $b = \sqrt{\hbar/2Nm\omega_z^2} \sim 10$ nm dephases with $\Delta\phi = 0.01 - 0.15°$ per cycle. While larger atom-surface distance reduces the anharmonicity, the coupling strength is also reduced, and longer interaction time is necessary to excite the cloud to a given amplitude. From an evaluation of Eq. 5.20 for the phase spread accumulated during the ecxitation to a given amplitude b as a function of the coupling strength parameter ϵ, one finds $\Delta\phi \propto 1/\epsilon$. This shows that small atom-surface distance is advantageous despite large anharmonicity.

5.8.2. Simulation of 1D Gross-Pitaevskii dynamics

In this section we use a numerical simulation of the Gross-Pitaevskii equation 2.26 in 1D to study the excitation of a BEC. In a first step, the ground state wave function for the repulsive gas in the potential $U(z) = U_m + U_s$ is found by imaginary time propagation. Then we perform time integration using a split-step Fourier method to

5.8 Simulation of cloud excitation

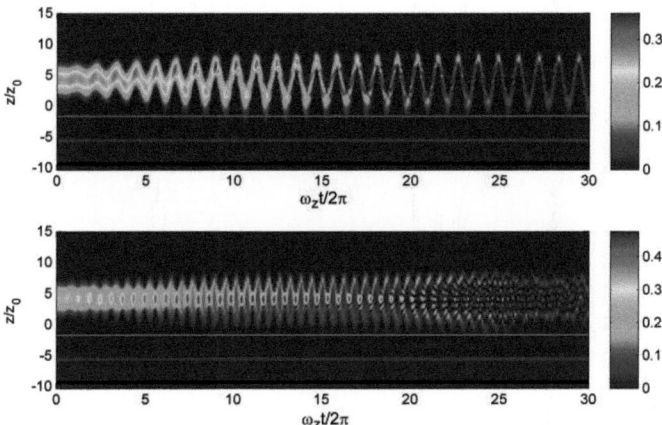

Figure 5.17.: Time evolution of a BEC with $N = 2000$ in a coupling trap with $\omega_z/2\pi = 10$ kHz (top) and $\omega_z/2\pi = 5$ kHz (bottom) at $d = 1.5$ μm for $t_h = 3$ ms. The cantilever amplitude is set to $a = 50$ nm (top) and $a = 70$ nm (bottom). Axes are in dimensionless units ($z_0 = \sqrt{\hbar/m\omega_z}$, the color scale indicates the normalized probability density $|\psi|^2$). Horizontal lines show the position of the barrier (grey), the matching layer (dark grey), and the cantilever surface (black).

obtain the temporal dynamics of the wave function in the time dependent potential $U(z,t)$. The code is adapted from [82, 270]. To avoid artificial Bragg scattering from the computational grid at high momentum in the attractive potential, it is necessary to introduce absorbing boundary conditions. We use the concept of a perfectly matched layer [271, 272] and implement an imaginary potential rather than an imaginary spatial coordinate. The idea is based on the fact that in a region, where a potential has a sizable imaginary contribution, matter waves are exponentially damped. The imaginary region is added to the potential in the form

$$U(z) = (U_m + U_{\text{CP}} + U_{\text{ad}}) \times \exp\left[-\frac{i\pi}{4}\left(1 - \tanh\frac{z - z_l}{d_z}\right)\right]. \quad (5.21)$$

This implements a smooth transition from a purely real potential for $z \gg z_l$ to an imaginary potential for $z \ll z_l$ within a matching layer of thickness d_z. For reasonably large d_z ($\sim 50-200$ nm), artificial reflections are very small and incident waves are completely absorbed. The position z_l is chosen such that the deBroglie wavelength remains much larger than the grid spacing, $\lambda_{dB}(z_l) = \hbar/\sqrt{2m(\mu_c - U(z_l))} \gg (z_{i+1} - z_i)$.

Figure 5.17 a shows the simulated cloud excitation for the c.o.m. mode at $\omega_m =$

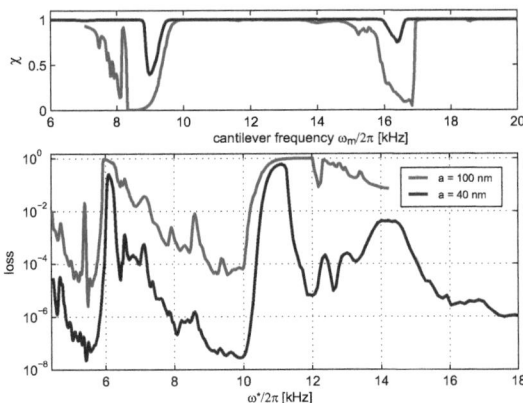

Figure 5.18.: BEC loss as a function of cantilever or trap frequency (trap parameters as in fig. 5.17). Upper panel: Fraction of remaining atoms χ as a function of ω_m for fixed $\omega_{z,0} = 2\pi \times 10$ kHz. Lower panel: Logarithmic plot of the contrast $C = 1 - \chi$. For comparison with experimental data we transform the axis to a frequency $\omega^* = \omega_{z,0}^2/\omega_m$ which is analog to the trap frequency in a scan with fixed ω_m and varied ω_z.

ω_z and the breathing mode at $\omega_m = 2\omega_z$. The dipole oscillation amplitude rises linearly in time until it saturates at an amplitude where the rim of the cloud reaches the trap barrier. Excitation of internal dynamics is visible from the modulation of the peak density. The breathing mode at twice the trap frequency increases also rather linearly in amplitude and saturates at a similar amplitude as the c.o.m. mode. For long times, dephasing leads to interference and finally to a decay of the breathing mode. The wave function temporarily shows only minor dynamics and is transformed into an excited state with ~ 4 nodes. With adequate timing, this could enable the preparation Fock states.

The atom number remaining after a coupling interval is simply evaluated by integrating over the density distribution. Simulating the coupling loss for different excitation frequencies yields a loss spectrum as shown in Fig. 5.18. For better control of the trap parameters, the cantilever frequency is varied rather than the trap frequency. This avoids e.g. unwanted changes in U_0, ϵ, or the contribution from tunneling, and thus permits to compare the strength of the resonances. The trap parameters are $\omega_{z,0}/2\pi = 10.0$ kHz and $d = 1.5$ µm. The shifted trap frequency for small amplitude oscillations is $\omega_z/2\pi = 9.434$ kHz. The c.o.m. and breathing mode resonance frequencies for medium oscillation amplitude are shifted by -450 Hz and -2488 Hz respectively. The resonances have opposite asymmetry, in agreement with

5.8 Simulation of cloud excitation

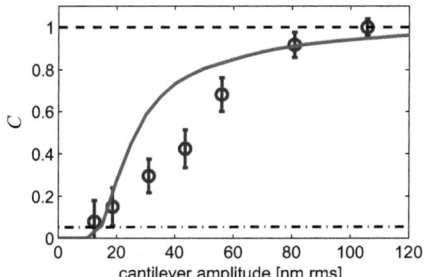

Figure 5.19.: Comparison of measured and simulated contrast as a function of cantilever amplitude for the c.o.m. mode. The parameters for the simulation are identical to those of the experiment without adjustments.

the classical simulation but deviating from experimental findings. The resonances of the 3D collective BEC modes are not reproduced (as expected), because the coupling to the other dimensions is not included. An interesting feature is the additional resonance in the spectrum for the stronger drive at $\omega_m \approx 8$ kHz. The cloud dynamics shows that several collective modes are excited at these subharmonic frequencies. This might explain the observed resonance lineshape in experiments for large drive and small atom-surface distance. Such modes do not exist in an harmonic trap and are also absent in the classical simulation for the anharmonic potential.

A central benefit of the simulation is to provide insight about the sensitivity limit of the used readout scheme. The simulation requires only the trap parameters and the cantilever amplitude as input, and has no free parameters to adjust. The only uncertainty comes from the error bars from the measurement of the trap frequency (see chapter 4.4.3). In Fig. 5.19, the simulated coupling loss as a function of the cantilever amplitude is compared to the measurement of chapter 5.11. Measurements and simulation agree surprisingly well, and especially the sensitivity predicted by the simulation is very close to our experimental findings. This is surprising, since we model a pure BEC in a 1D trap, whereas the experiment deals with partially condensed clouds with thermal fraction in a 3D trap. Possible effects could be e.g. nonlinear coupling to other trap axes, variation of ω_z along the trap axis, dependence of the barrier height on the axial distance to the trap center, and damping of condensate motion by the thermal cloud. However, it seems that these effects play a minor role for the loss.

6. Outlook

The experiments reported in this thesis represent a first step in the coupling of ultracold atoms to mechanical resonators. One of the long term motivations for this research is to achieve such a coupling on the quantum level. In this outlook I discuss three different scenarios for atom-resonator coupling, all of them having the potential to achieve observable back action of the atoms on the resonator. We furthermore show that for cryogenic temperatures and high mechanical quality factors, the strong coupling regime can be reached. In this regime, the presence of a single excitation quantum in one of the two subsystems leads to an energy exchange rate faster than all dissipation or decoherence rates, enabling coherent coupling on the quantum level. This is the key ingredient for the realization of hybrid quantum systems that would allow one to create atom-resonator entanglement, quantum state transfer, and quantum control of mechanical force sensors.

6.1. Coupling BECs to a carbon nanotube

Coupling ultracold atoms and mechanical oscillators via surface forces has one central advantage: The employed coupling force is of fundamental origin, and functionalization of the oscillator, e.g. by the fabrication of mirrors, magnets or electrodes on the resonator, is not required. This is in contrast to most of the theoretical proposals considering atom-resonator coupling so far [35, 36, 37, 38, 39, 40, 41, 42, 43, 44, 45, 46, 47, 48]. Coupling via surface forces could thus be used to couple atoms to molecular-scale oscillators. The importance of this fact becomes clear when recalling the coupling strength derived in chapter 5.1.3. We have found that the quantum mechanical coupling rate between the motion of a single atom of mass m and the motion of a mechanical resonator of mass M and resonance frequency ω_m scales as

$$g_0 = \frac{\epsilon \omega_m}{2} \sqrt{m/M_{\text{eff}}}. \qquad (6.1)$$

In our case, $\epsilon = \delta z_t / a$ describes the ratio between resonator amplitude a and trap position shift amplitude δz_t. The equation shows that for a large coupling rate, high frequency and extreme-low mass oscillators are desirable. The easiest way to increase the coupling, by choosing an oscillator with high eigenfrequency, is limited by the achievable atomic trap frequencies. For a cloud of atoms, trap frequency dependent loss like inelastic three-body collisions imposes an even harder limit (see chapter

2.5.1 and 4.4). Moreover, just increasing the eigenfrequency is not necessarily a solution, as all dissipative rates will also in most cases increase proportional to the eigenfrequency (if we assume fixed Q). A more beneficial direction is thus the employment of oscillators with ultra-low mass.

Due to their ultimately low mass and the high achievable quality factor, single wall carbon nanotubes (SWCNT) are particularly promising in this context. Techniques for controlled growth enable the realization of suitable geometries, e.g. several microns long, single clamped nanotubes [273, 274]. Alternatively, also double clamped nanotubes might be interesting, as sensitive on chip readout and actuation schemes exist [275, 276, 277, 278, 279]. A central point is that mechanical dissipation in SWCNTs can be very low, e.g. a quality factor of $Q > 10^5$ was achieved recently [202]. Alternatively, SiC nanowires have been shown to also reach exceptional Q-factor (1.6×10^5) [201] and can be fabricated in the desired frequency range. Finally, suspended graphene sheets have been recently studied as mechanical resonators [280, 203], and Q-factors up to 10^4 have been demonstrated at low temperature.

SWCNTs are typically produced with a diameter in the range of $d_{\mathrm{CNT}} = 1-4$ nm. Their Youngs modulus amounts to $E = 1$ TPa, comparable to the value of diamond. Their mass can be directly determined from the surface mass density of graphene, $\rho_{2D} = 7.7 \times 10^{-7}$ kg/m^2, giving a length mass density $\rho_{1D} = \pi d_{\mathrm{CNT}} \rho_{2D}$, and a typical value is $\rho_{1D} = 5 \times 10^{-21}$ kg/μm.

A nanotube has a fundamental resonance frequency [276]

$$\omega_m = k_1^2 \sqrt{\frac{EI}{\rho_{1D}}} \tag{6.2}$$

with $I = \pi d^4/64$ the moment of inertia and $k_1 = (1.875, 4.730)/l$ for a single and double clamped nanotube respectively with suspended length l. For a SWCNT with $d_{\mathrm{CNT}} = 1.5$ nm and $l = 15$ μm we obtain $\omega_m = 20$ kHz and a mass of $M = 6 \times 10^{-20}$ kg, or effective mass $M_{\mathrm{eff}} = 0.24\, M$. This leads to a room temperature thermal motion amplitude of $a_{th} = \sqrt{k_B T / M_{\mathrm{eff}} \omega_m^2} = 4$ μm, much larger than typical atom-surface distances in our experiment. Even more impressive, the ground state amplitude is almost macroscopic,

$$a_{qm} = \sqrt{\frac{\hbar}{2 M_{\mathrm{eff}} \omega_m}} = 0.2 \text{ nm}. \tag{6.3}$$

These numbers are about four orders of magnitude larger than those of our cantilever. When the readout scheme of direct imaging of the cloud excitation (see chapter 5.6) is employed, the expected sensitivity should permit measurements close to the ground state of such a nanotube.

However, one central concern is whether the strength of the surface potential of such nanoscale objects is sufficient for coupling experiments. The Casimir-Polder

6.1 Coupling BECs to a carbon nanotube

potential of a SWCNT was calculated by Fermani *et al.* [281]. From an approximative reproduction of the surface potential in the distance range $d = 50 - 250$ nm as shown in this article, we infer a potential $U_{\rm CP,SWCNT} \approx 0.06 \times U_{\rm CP}$, a factor 17 smaller than the CP potential of a bulk conductor. A trap with $\omega_z/2\pi = 20$ kHz at a distance $d = 600$ nm from the nanotube has a trap depth $U_0 = 80$ kHz, similar as in our coupling measurements. For this trap we calculate a coupling strength parameter $\epsilon = 5 \times 10^{-3}$, a factor $\beta = 10$ smaller than that expected for the CP potential of a perfect conductor and a factor $\beta = 30$ weaker than measured with the stronger surface potential above the metallized side of our cantilever. By approaching the nanotube closer and using a trap with smaller U_0, this can be partially compensated, e.g. at $d = 480$ nm we have $U_0 = 34$ kHz and $\epsilon = 0.02$.

Alternatively, stronger coupling could be realized by static charging of the CNT. Due to the small atom-surface distance, the field $E \sim V/d_{\rm CNT}$ and also the field gradient will be enhanced dramatically. Thereby, a potential much stronger than the CP potential could be created, and coupling strength parameters of $\epsilon \sim 0.3$ might be possible. Figure 6.1 shows the coupling strength parameter ϵ as a function of the strength of the surface potential. The observed ϵ at the both sides of the cantilever used in our experiments is shown together with the expected value for a SWCNT. Since CNTs can be fabricated in metallic or semiconducting configuration, charging does not require fabrication on the nanotube, and electric contacting is a standard technique that is compatible with high quality factors [202].

Finally, coupling schemes that do not rely on surface forces could be employed. One possibility is to use a doubly clamped, suspended SWCNT as current carrying wire to generate a magnetic trap. Several papers have theoretically investigated this situation [152, 153, 281] and studied its feasibility for atom trapping. In this configuration, mechanical oscillations of the nanotube translate directly into trap oscillations and provide a coupling strength parameter $\epsilon = 1$. In combination with compensating fields with convex curvature ($U'' > 0$), even values of $\epsilon > 1$ could be possible. A different possibility is to couple to an internal degree of freedom of the atoms. A constant current sent through an oscillating nanotube will create an oscillating magnetic field at the position of the atoms that is proportional to the field gradient and the nanotube oscillation amplitude. If the magnetic field oscillations are resonant with a Zeeman transition of the atoms, observable spin-flip transitions are induced. This coupling mechanism is discussed in the next chapter for the case of a nanomechanical resonator with a magnetic tip.

An independent readout of the nanotube is highly desirable, e.g. to observe the back action directly. Using capacitive readout for a doubly clamped, semiconducting SWCNT, Hüttel *et al.* [202] have achieved a resolution of ~ 0.3 nm for a 300 MHz oscillation. An alternative readout scheme that is capable to sense single clamped nanoscale oscillators relies on dispersive or absorptive interaction of the oscillator with an optical cavity mode [282, 76]. With a fiber based optical micro-cavity devel-

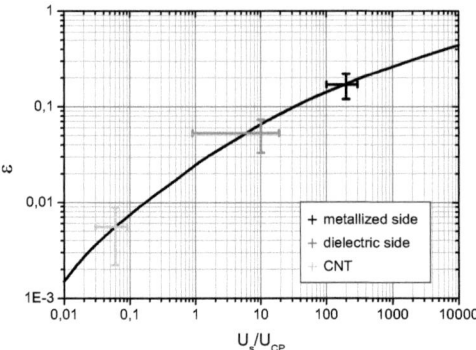

Figure 6.1.: Variation of the coupling strength parameter $\epsilon = \delta z_t/a$ as a function of surface potential strength normalized to the CP potential of a perfect conductor, calculated for a trap depth of $U_0 = 80$ kHz (solid line). In addition, the evaluated ϵ above the metallized and dielectric side of our cantilever and the expected value for a SWCNT are shown.

oped in our group [79], the thermal motion of a carbon nanorod with fundamental resonance frequency of 470 kHz could be measured with a sensitivity of 0.2 pm/$\sqrt{\text{Hz}}$ [76]. This technique might also allow one to cool the oscillators motion.

Strong coupling parameters

We now estimate the achievable coupling parameters for a SWCNT coupled either to a small BEC or to a single atom. For an atom cloud, three-body collisional loss is the dominating dissipative process, which also restricts the choice of the trap frequency (see chapter 2.5.1). For $N = 500$ atoms in an elongated trap with aspect ratio $\omega_z/\omega_x = 25$ that is resonant with the CNT considered above with frequency $\omega_m = 20$ kHz, we obtain an atomic loss rate of $\gamma_a = 2\pi \times 13$ Hz, and for a moderate quality factor of $Q = 10^4$ we have a mechanical damping rate $\kappa = \omega_m/2Q = 2\pi \times 2$ Hz. This is to be compared to the single-phonon single-atom coupling rate, which amounts to $g_0 = 2\pi\epsilon \times 35$ Hz, and for an ensemble of atoms, the coupling is collectively enhanced according to $g_N = g_0\sqrt{N}$. On the other hand, for a single atom, the only dissipative processes are collisions with the background gas, technical current noise induced heating, or Johnson noise induced spin-flip loss (see chapter 2.5). These effects are small or can be controlled, and we assume an atomic loss rate $\gamma_a = 2\pi \times 1$ Hz. Here, the trap frequency can be made as large as possible, and in a magnetic microtrap, frequencies of $\omega_z = 250$ kHz should be possible. For a SWCNT with $l = 4.25$ µm

6.2 Magnetic coupling of a BEC to a nanomechanical resonator

such that $\omega_m = \omega_z$ we find a coupling rate $g_0 = 2\pi\epsilon \times 800$ Hz, and for the same quality factor as above we obtain $\kappa = 2\pi \times 12$ Hz. Comparing the coherent coupling $g = \{g_0, g_N\}$ and dissipative rates for both cases we obtain

$$(g, \kappa, \gamma_a) = 2\pi \times \begin{cases} (\epsilon \times 780,\ 1, 13)\ \text{Hz} & \text{collective} \\ (\epsilon \times 800,\ 12, 1)\ \text{Hz} & \text{single atom.} \end{cases} \quad (6.4)$$

This is sufficient to enter the strong coupling limit both for a single atom and collectively for an atom cloud with comparable rates. Furthermore, the strong coupling condition is met for a large range of coupling strength parameters, down to $\epsilon \sim 0.02$. While the overall rates are quite low, we would like to emphasize that the relative coupling strength $g_0/\omega_m = \epsilon \times 3 \times 10^{-3}$ is enormous. The strongest relative coupling strength so far was demonstrated in a microwave cavity dispersively coupled to a superconducting qubit, where $g_0/\omega_{cav} = 105\ \text{MHz}/5.7\ \text{GHz} = 2 \times 10^{-2}$ was obtained [283]. For comparison, the strongest optical cavity QED coupling so far was achieved with fiber based micro-cavities [79] and amounts to $g_0/\omega_{cav} = 215\ \text{MHz}/380\ \text{THz} = 6 \times 10^{-7}$ [75].

The low overall rates constitute a setting where the oscillator environment has strong impact. While thermal excitations of the mode under study are easily frozen out in microwave or optical cavity QED experiments, this is challenging for low frequency mechanical oscillators. A low bath occupation is essential, as the decoherence rate of an oscillator coupled to a bath scales with the mean thermal occupation $n_{\text{th}} \simeq k_B T/\hbar\omega_m$ (see chapter 3.3.1),

$$\gamma_{\text{dec}} = n_{\text{th}}\kappa = \frac{k_B T}{\hbar Q}. \quad (6.5)$$

The requirements for the achievement of useful quantum state lifetimes are ambitious but within reach, e.g. for $1/\gamma_{\text{dec}} = 10$ ms one needs $T = 10$ mK and $Q = 10^7$.

We note that several cryogenic atom chip experiments have already been demonstrated [284, 285, 286, 287, 288, 289, 290], and BEC has been achieved [287]. Most of the experiments are operated at the liquid He base temperature of ~ 4 K, and superconducting trapping wires are employed to avoid heat load. This should enable the operation of magnetic traps also at much lower temperatures.

6.2. Magnetic coupling of a BEC to a nanomechanical resonator

In this chapter I summarize a theoretical proposal where we study a magnetic coupling between a nanoscale mechanical resonator and the internal state of atoms. We propose an experimental setting for the realization of this system and discuss the potential of reaching the strong coupling limit. This work was published in [39] and summarized in [82].

Coupling to internal states has the advantage that high frequency mechanical resonators can be employed, as the coupling now is to be resonant e.g. with Zeeman transitions at typically MHz frequencies. Additionally, the internal levels can be chosen and modified (e.g. by state selective microwave level dressing [9]) such that the atoms act as effective two-level systems. This is important for the generation and readout of quantum states, because in a two-level system, Fock states and superposition states can be prepared by classically driving Rabi oscillations. Furthermore, the quantum control of internal states is more advanced than that for collective motion, at least in magnetic traps.

Coupling mechanism

We consider the setting of a nanomagnet on the tip of a nanomechanical cantilever whose magnetic field couples to the atomic spin and drives Zeeman transitions. Figure 6.2 shows the situation with the involved transitions. The nanomagnet transduces mechanical oscillations of the cantilever into magnetic field oscillations at the position of the atoms. The amplitude of the field oscillations is linked to the cantilever oscillation $a(t)$ via the field gradient G_m,

$$\boldsymbol{B}_r(t) = a(t) G_m \boldsymbol{e}_x. \tag{6.6}$$

The interaction of the atomic spin with $\boldsymbol{B}_r(t)$ is described by the Zeeman Hamiltonian $H_Z = -\boldsymbol{\mu} \cdot \boldsymbol{B}_r(t) = \mu_B g_F F_x G_m a(t)$, where $\boldsymbol{\mu} = -\mu_B g_F \boldsymbol{F}$ is the magnetic moment and $\boldsymbol{F}$ the atomic spin. The field direction is chosen orthogonal to the field in the trap center, such that the oscillations can drive spin-flip transitions between neighboring Zeeman sublevels (see figure 6.2(b)). The spin-flip transition rate becomes sizable when the field oscillations at frequency ω_m are resonant with the Larmor frequency $\omega_L = \mu_B |g_F| B_0 / \hbar$ of the atoms in the trap, set by the value of the overall magnetic field B_0 in the trap center. Since B_0 can be easily varied experimentally, the resonance condition $\omega_m = \omega_L$ can be fulfilled for a large range of cantilever frequencies. Furthermore, by rapidly changing B_0 the coupling can be switched on and off.

For detection of the coupling induced spin-flip rate one can either rely on transitions to untrapped states, as e.g. given for the transition $|1, -1\rangle \to |1, 0\rangle$, or one can implement state-selective detection to monitor the population of e.g. the state $|2, 2\rangle$ coupled to $|2, 1\rangle$.

In the case of a transition to an untrapped state, e.g. for a BEC in state $|1, -1\rangle$ coupled to $|1, 0\rangle$, the outcoupling rate Γ_r from the trap has to account for the energy broadening of the cloud in the trap, which directly translates into a broadening of the Larmor frequency. This situation is known from atom laser experiments [291, 292], and the coupling to the untrapped state with a Rabi frequency $\Omega_R = \mu_B G_m a / \sqrt{8} \hbar$

6.2 Magnetic coupling of a BEC to a nanomechanical resonator

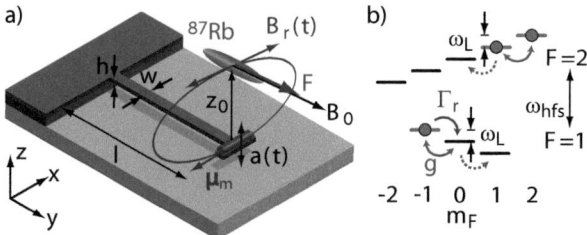

Figure 6.2.: (a) Schematic setting of the magnetic coupling of a nanomechanical resonator to a BEC. A nanomagnet on the tip of the cantilever transduces mechanical motion into magnetic field oscillations. (b) Involved ground state levels. Magnetic field oscillations at the Larmor frequency ω_L lead to a coupling g between different Zeeman sublevels.

leads to a loss rate

$$\Gamma_r = \frac{15\pi}{8} \frac{\hbar \Omega_R^2}{\mu_c} (r_c - r_c^3) \qquad (6.7)$$

for a BEC in the Thomas-Fermi limit and for weak coupling $\hbar\Omega_R \ll \mu_c$. Here, $r_c = \sqrt{\hbar\delta/\mu_c}$, and output coupling takes place on a thin ellipsoidal resonance shell with main axes $r_i = r_c R_{TF,i}$.

Simulation and chip layout

We perform a detailed simulation of the system in order to optimize a possible experimental geometry for the realization of the system. Figure 6.3 (a) shows an optimized setting of current carrying wires, an integrated nanoresonator, and coupling as well as compensation magnets. The latter have to be introduced to reduce the distorting effect of the coupling magnet on the trap. Long compensation magnets on the substrate with small gaps to the coupling magnet on the resonator bend away magnetic field lines, such that the static field gradients are suppressed while the field oscillation amplitude remains unaffected. Figure 6.3 (b) shows the resulting potential seen by the atoms. It includes the magnetic trapping potential generated by the wires shown in figure 6.3 (a) together with homogeneous bias fields B_b, the potential due to the coupling and compensation magnets, the Casimir-Polder surface potential, and gravity. The effect of the magnets is visible as a strong repulsive deformation close to $y = 0$ μm.

The trap is optimized to maximize the ratio between Γ_r and three-body collisional loss γ_{3b}, and we find optimal values as quoted in figure 6.3. The resonator is a Si cantilever with dimensions $(l, w, t) = (7.0, 0.2, 0.1)$ μm, leading to a resonance

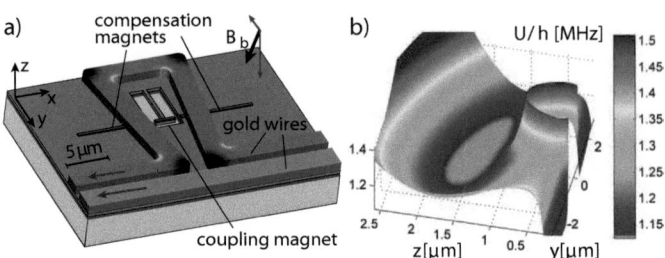

Figure 6.3.: (a) Proposed layout. Currents in 2 µm wide gold wires of typically 4 mA together with a homogeneous field $B_b = (-0.1, 4.2, 6)$ G create a magnetic trap with trap frequency $\omega_{x,y,z}/2\pi = (8.9, 1.2, 9.7)$ kHz at $(y_0, z_0) = (0.0, 1.5)$ µm. The wire color indicates the current density obtained from a finite elements simulation. The coupling magnet is located at the tip of the cantilever at $(y, z) = (0, 0)$ µm, and two compensation magnets are aligned in a row with it to reduce trap distortion. (b) 2D cut through the potential energy for the atoms in the y, z-plane. Included are the magnetic potential from the wires, the homogeneous field B_b, the stray field of the three magnets, the Casimir-Polder surface potential and gravity. The orange colored area indicates the extension of a BEC of 1000 atoms.

frequency of $\omega_m/2\pi = 1.12$ MHz and an effective mass $M_{\text{eff}} = 3 \times 10^{-16}$ kg. The nanomagnet for coupling is chosen to be a single domain Co island of dimensions $(l_m, w_m, t_m) = (1.3, 0.2, 0.08)$ µm, the two compensation magnets have same cross section and 5 µm length, and the gap between the magnets is 200 nm.

The simulation predicts that the coupling rate Γ_r for a thermally driven cantilever at room temperature can be made much larger than the dissipative rates γ_a for the atoms and $\kappa = \omega_m/2Q$ for the nanoresonator. For experimentally realistic quality factors $Q = 5000$ and a loss rate dominated by three-body collisions calculated for 1000 atoms in a trap as given in figure 6.3, we obtain

$$(\Gamma_r, \gamma_a, \kappa) = (2.1, 0.02, 0.74) \text{ kHz}. \tag{6.8}$$

This achieves the condition for coherent coupling and permits to use the BEC as a direct probe for the thermal fluctuations of the cantilever amplitude $a(t)$. For such a measurement, the BEC in state $|1, -1\rangle$ is coupled to the cantilever for a time $\tau \ll \kappa^{-1}$ and the remaining atom number $N(a, \tau) = N \exp[-\Gamma_r(a)\tau]$ is measured. In a single shot of the experiment, the cantilever has a well defined and approximately constant amplitude $a(\tau)$. Repeating the experiment samples different cantilever amplitudes and results in fluctuations of $N(a, \tau)$. Because $\Gamma_r \propto a^2 \propto n$, the fluctuations reflect the distribution of the phonon number n, with the mean value

6.2 Magnetic coupling of a BEC to a nanomechanical resonator

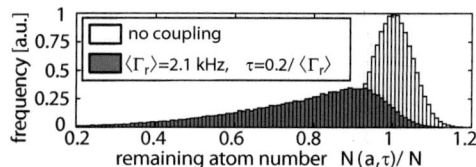

Figure 6.4.: Simulation of the statistics of coupling loss measurements for a thermally driven cantilever. The histogram shows the fraction of remaining atoms after a coupling time $\tau = 0.2/\langle \Gamma_r \rangle$ including background loss. We have assumed ±5% fluctuations in atom number due to the preparation.

$n_{th} = [\exp(\hbar\omega_m/k_BT) - 1]^{-1}$ defining $\langle \Gamma_r \rangle$. Figure 6.4 shows a simulated histogram for a mean coupling rate $\langle \Gamma_r \rangle = 2.1$ kHz.

The geometry and the parameters of the simulation have been optimized in close interplay with the development of a fabrication process for the system. This is work carried out by Stephan Camerer in collaboration with the group of Jörg P. Kotthaus.

Mechanical cavity QED

At room temperature, the phonon occupation of the cantilever is macroscopic ($n_{th} \sim 5 \times 10^6$), permitting a classical description of the coupling, and the back action of the atoms on the resonator is negligible. In contrast, for low temperatures, where the atom number is comparable or larger than the mean phonon occupation, $N \sim n_{th}$, back action can become significant and might be used to manipulate the resonator on the quantum level. In this regime, the system has to be described quantum mechanically. A system of N two-level systems coupled to a harmonic oscillator can be described by the Tavis-Cummings Hamiltonian [293], a multi-particle version of the broadly known Jaynes-Cummings Hamiltonian for a single two-level system. The coupling is described by a single-atom single-phonon coupling constant $g_0 = \mu_B G_m a_{qm}/\sqrt{8}\hbar$, and for an ensemble of N identically coupled atoms, the coupling is again collectively enhanced and leads to a coupling constant $g_N = \sqrt{N}g_0$.

If the coherent coupling rate can be made larger than all dissipative rates, $g_N > (\kappa, \gamma)$ where γ summarizes all atomic loss rates, an equivalent to the strong coupling limit of cavity QED [30, 294, 295, 296] is reached.

A simulation for a cloud of $N = 10^4$ atoms in a trap with $\omega_{ho} = 2.9$ kHz at a distance $d = 2.0$ μm from a 1.1 MHz resonator with $Q = 10^5$ predicts the achievement of the collective strong coupling limit. Alternatively, a smaller resonator with $\omega_m/2\pi = 2.8$ MHz coupled to a single atom in a tight trap with $\omega_{ho} = 250$ kHz at distance $d = 250$ nm would enter the single atom strong coupling limit. Summarized, the system could reach coupling parameters

$$(g, \kappa, \gamma) = 2\pi \times \begin{cases} (21, \ 5, 10) \text{ Hz} & \text{collective} \\ (62, 14, \ 1) \text{ Hz} & \text{single atom.} \end{cases} \quad (6.9)$$

To prepare the resonator in the ground state, a temperature $T < 100$ μK is needed (e.g. possible in a nuclear demagnetization cryostat [244]), while the regime $n_{th} \sim N$, where collective effects should become visible, is reached already for 50 mK, a temperature achievable in a dilution refrigerator. This suggests that the strong coupling regime is accessible for low bath temperatures and a high quality factor of the cantilever.

6.3. Optical coupling of ultracold atoms to mechanical resonators

So far we have discussed coupling mechanisms that rely on close approach of the atoms to a mechanical resonator. They are based on forces generated by the mechanical resonator that decay over a few micrometer distance. This requires exceptional control both of the position of the cloud with respect to the oscillator, and of the temperature of the atoms, i.e. Bose-Einstein condensates to achieve the minimum possible extension of the cloud.

In this section we briefly introduce a coupling scheme that relies on the long distance coupling between a mechanical resonator and laser cooled atoms.

Consider a dipole trap laser that is retro reflected on a mechanical resonator to form a 1D standing wave, suitable to trap laser cooled atoms. Movements of the resonator will move the antinodes of the standing wave, such that the resonator oscillations are directly transferred to an oscillation of the trapping potential of the atoms. When the cantilever oscillation frequency is resonant with the trap frequency of the atoms along the lattice direction, resonant excitation of the collective c.o.m. mode of the atoms leads to an energy transfer from the resonator to the atom cloud.

Furthermore, there is also a back action of the atoms on the oscillator: The restoring force that holds the atom at an anti-node of the standing wave is provided by coherent redistribution of photons between the two k-vectors that make up the lattice. E.g. if an atom is displaced to the right of an anti-node, it will scatter photons from the beam coming from the right to the beam that runs to the right, such that the photon recoils will push it back to the center. This redistribution leads to a modulation of the power of the beam impinging on the resonator, which results in a modulation of the radiation pressure force on the resonator due to the reflection of the beam. When the atoms are excited by the resonator to collective motion, there is a fixed phase relation between atomic and resonator motion, and the resulting radiation pressure modulation at the resonator will have just the right

6.3 Optical coupling of ultracold atoms to mechanical resonators

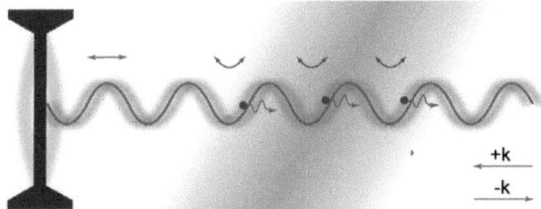

Figure 6.5.: Schematic coupling mechanism: A lattice laser is retro reflected on a mechanical resonator to form a standing wave dipole trap. Oscillations of the resonator shake the lattice and excite c.o.m. motion of atoms trapped in the lattice. Laser cooling of the atoms provides sympathetic cooling of the resonator.

delay to damp the oscillations. To avoid heating of the atoms out of the trap due to the excitation, one can apply laser cooling to the atoms and thereby continuously extract the gained energy of the atoms. The motion of the mechanical resonator can in this way be sympathetically cooled via laser cooling of the atoms. Figure 6.5 depicts the situation.

This scheme has several remarkable advantages. Most prominently, the lattice provides a coupling that can bridge large distances (several hundred meter), limited only by retardation of the light field and by the frequency stability or coherence of the laser. This relaxes technical effort considerably, as the mechanical resonator does not have to be integrated into a cold atom UHV environment. It also simplifies the implementation of a cryogenic environment for the resonator. On the atomic side it is sufficient to prepare laser cooled atoms instead of BECs, and no positioning is necessary. This simplifies and shortens the experimental cycle considerably. Furthermore, laser cooling of the trapped atoms can be realized in several ways. E.g. optical molasses cooling in the lattice already provides sufficient temperature control. Finally, the coupling strength scales with the atom number and can be made large by loading a large ensemble of atoms into the lattice. Altogether, this promises to observe back action effects already at room temperature in a fairly simple setup.

We started to realize this experiment and initiated a collaboration with the group of Peter Zoller on the theoretical description of the coupling [47].

6.4. Conclusion

In this thesis I introduce a new type of experiment that interfaces solid state physics and quantum optics. The motivation for such research is manifold, amongst it the exploration of atom-solid state interfaces for quantum information tasks, the extension of quantum control to more complex or even macroscopic systems, and the achievement of quantum limited sensitivity e.g. for position or force sensing.

I report experimental results on coupling a Bose-Einstein condensate to the driven motion of a micromechanical resonator via surface forces. The experiment is a first demonstration of the direct and controlled coupling between a single degree of freedom of a solid state system and a gas of ultracold atoms. Due to the short-range nature of the coupling force, high spatial control over the atoms is necessary. We demonstrate such high control over the atoms at small distance from the resonator surface, and quantify the effects that limit experiments under such conditions. We observe a clear signature of the coupling by detecting increased atom loss. In this way, the atoms can be used to sense mechanical oscillations. We study the spatial distance range, the temporal evolution, and the resonant character of the coupling. We observe sharp resonances in the frequency spectrum of the atomic response to the coupling, indicating coherent excitation of collective modes of the condensate. Backed by numerical simulations we find a consistent picture of the coupling dynamics and an explanation for the observed sensitivity.

The reported experiment is a first step on the way to realize a coherent link between ultracold atoms and mechanical resonators. Yet, considerable challenges have to be overcome to achieve the experimental conditions required for a coupling on the quantum level. The most fundamental issue is to achieve a coherent coupling rate that exceeds the decoherence rate of the resonator, and the technically most demanding one is the implementation of such an experiment in a cryogenic setup.

To engineer a stronger interaction it is advisable to explore various coupling schemes. The three proposals reported in the outlook have the potential to show back action of the atoms onto the resonator, the next important step on the way to a hybrid quantum system. We furthermore show that the regime of mechanical strong coupling can be achieved, limited essentially by the achievable mechanical quality factor and the bath temperature. Recent cryogenic experiments with cold atoms or BECs show that it is indeed possible to combine atom trapping and cryogenics. In conjunction with the ongoing improvements in the mechanical properties, the readout, and the control of micro- and nanomechanical resonators, it seems that the era of quantum *mechanics* is about to come.

A. Fundamental constants and Rubidium Data

Fundamental constants

Speed of light	c	$2.997\ 924\ 58 \times 10^8$ m/s
Planck's constant	h	$6.626\ 068\ 76(52) \times 10^{-34}$ J s
Bohr magneton	μ_B	$1.399\ 624\ 624(56)$ MHz/G
Bohr radius	a_0	$0.529\ 177\ 208\ 3(19)\ 10^{-10}$ m
Boltzmann's constant	k_B	$1.380\ 650\ 3(24) \times 10^{-23}$ J/K

Rb physical properties

Atomic number	Z	37
Total nucleons	$Z+N$	87
Relative abundance		27.8 %
Atomic mass	m	$1.443\ 160\ 60(11) \times 10^{25}$ kg
Melting point	T_M	$39.31°$C
Vapor pressure at 25°C	P_V	4.0×10^{-7} mbar
Nuclear spin	I	3/2

$5^2S_{1/2}$ ground state properties

Hyperfine splitting	E_{hfs}	$h \cdot 6.83468261090429(9)$ GHz
Electron spin g-factor	g_J	$2.002\ 331\ 13(20)$
Nuclear spin g-factor	g_I	$-0.000\ 995\ 141\ 4(10)$
Ground state polarizability	α_0	$h \cdot 0.0794(16)$ Hz/(V/cm)^2

D_2 transition ($5^2S_{1/2} \to 2^2P_{3/2}$) optical properties

Transition frequency	ω_0	$2\pi \cdot 384.230\ 484\ 468\ 5(62)$ THz
Vacuum wavelength	λ	$780.241\ 209\ 686(13)$ nm
Natural line width	Γ	$2\pi \times 6.065(9)$ MHz
Saturation intensity	I_s	$1.669(2)$ mW/cm^2
Recoil energy	$E_r = 2\pi^2\hbar^2/\lambda^2 m$	$h \cdot 3.7710$ kHz
Doppler temperature	$T_D = \hbar\Gamma/2k_B$	146 μK

Data taken from [86].

B. Fast trap ramping

To transport the atoms to the cantilever we use ramps where the trap frequency and the position are changed simultaneously at a rate close to the limit of adiabaticity. This is important because three-body collisional loss limits the lifetime in the compressed trap to ~ 17 ms and technical heating leads to thermalization within 3 ms when the cloud is away from the surface. To avoid cloud excitation we use Blackman pulse shaped ramps. They define the velocity of the trap center according to

$$v(t, t_r) = \frac{1}{t_r}\left(1 - \frac{25}{21}\cos(2\pi\frac{t}{t_R}) + \frac{4}{21}\cos(4\pi\frac{t}{t_r})\right). \tag{B.1}$$

To minimize changes in the power dissipation on the chip which would affect the eigenfrequency of the cantilever (see chapter 4.1.2), we ramp the bias field rather than the wire current. The used power supply (FUG 12A 15V) allows ramp times $t_r > 0.7$ ms, but slight deviations from the set ramp as well as asymmetries show up for $t_r < 1.5$ ms. Furthermore, we apply the ramp to the bias field $B_{b,y}$, which translates into position according to $z_t \approx (\mu_0/2\pi)I/B_{b,y}$. As the relative change during the ramp $\Delta B_{b,y}/B_{b,y} \approx 30\%$ is not small, the nonlinear dependence $z_t(B_{b,y})$ will cause a slightly distorted position ramp. The ramp time is chosen such that the change in trap frequency fullfills the adiabaticity criterion

$$\frac{\partial \omega_z}{\partial t} \ll \omega_z^2 \tag{B.2}$$

at all times during the ramp. Fig. B.1 shows a calculation for a ramp from the condensation trap to a coupling trap with final trap frequency $\omega_z/2\pi = 10$ kHz. The lower panel of (a) shows the criterion B.2.

A condition to describe how well the cloud remains at the trap center can be formulated by demanding forces during the ramp to be smaller than the restoring force for a cloud displaced by one ground state extension $b_{\text{qm}} = \sqrt{\hbar/2m\omega_z}$,

$$\frac{\partial^2 z}{\partial t^2} < \omega_z^2 b_{\text{qm}}. \tag{B.3}$$

This is not always fulfilled for the used ramps. Excursions of the center of mass with respect to the trap center amount to $\sim 2 \times b_{\text{qm}} = 300$ nm. However, due to the symmetry of the ramp, the cloud should be at the trap center at the end of the pulse again. The upper panel of Fig. B.1 b) shows the velocity of the trap center

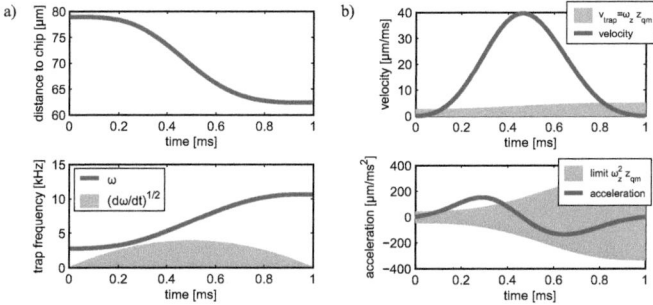

Figure B.1.: Trap parameters during ramps to the surface.

and for comparison the rms velocity spread of the ground state $v_{\rm qm} = \omega_z b_{\rm qm}$. In the lower panel of Fig. B.1 (b) the condition B.3 is shown.

The adiabaticity of the ramp is verified experimentally (see chapter 5.2 for details).

C. Manoeuvre around the cantilever

For measurements on the dielectric back side of the cantilever it is desirable to have conditions comparable to the measurements on the front side. This requires a non-trivial transport around the cantilever and as a consequence a modified atom preparation. We find it advantageous to orbit the cantilever along the axial direction of the trap (the x-axis) and use a neighbouring dimple wire ("shift wire") to axially move the trap.

The ramping involves several phases and a relaxed trap frequency along the axial direction during one stage, such that the ramping could easily cause excitations or even destroy a BEC. We therefore prepare a cloud with a temperature of $\sim 200\mu K$ with the second RF evaporation ramp (RFC) stopped 2.6 MHz above the trap bottom (other parameters as in RFC B, see table 4.3 in chapter 4.3). After bypassing the cantilever, a BEC is then prepared a in a trap at 15 μm distance from the back

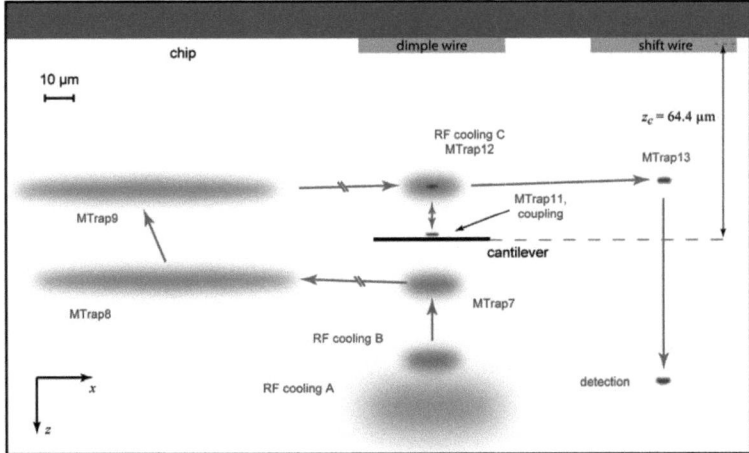

Figure C.1.: Schematic geometry of the manoeuvre in the x, z-plane. The goal is to prepare BECs on the back side of the cantilever and perform measurements under comparable conditions as on the front side.

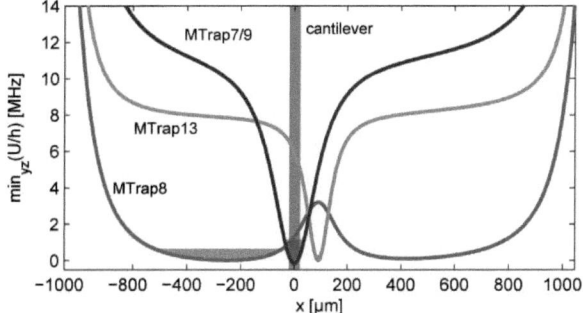

Figure C.2.: Potential minimum along the x-axis. Dimple trap before (MTrap7) and after (MTrap9) passing the cantilever. Ioffe trap with repulsive barrier during the bypass of the cantilever (MTrap8). A tight dimple trap above the shift wire is used for fast ramping into the detection trap (MTrap13).

side of the cantilever. For imaging, the cantilever has to be bypassed again. We want to minimize background and collisional loss after coupling measurements, and we use a different path with higher trap frequencies to allow for a quick withdrawal from the chip. Figure C.1 schematically shows the geometry of the situation. The axial trapping potentials during the important phases are displayed in Figure C.2.

The following steps comprise the manoeuvre:

- To axially shift the cloud away from the cantilever we ramp down the dimple confinement and use the shift wire to produce a repulsive bump along x, while z_t remains approximately constant at $z_t = 80$ μm (MTrap8). This shifts the cloud by 250 μm along x, but also relaxes the axial confinement to $\omega_x/2\pi = 43$ Hz, giving a cloud radius of $R_x = 65$ μm.

- This trap is then ramped towards the chip to a distance $z_t \sim 51$ μm from the chip, thereby bypassing the cantilever.

- After transforming back into the dimple at the cantilever we perform RF evaporation (RFCoolingC) and make coupling measurements with parameters very similar to those on the other side of the cantilever (MTrap11).

- We then ramp back into the condensation trap (MTrap12) which we transform into a dimple formed by the shift wire to by-pass the cantilever for detection. However, during crossfading between the two dimples, the intermediate potential forms a double well, and the atoms "fall down" a potential step of a few hundreds of kHz when the well centered at the cantilever vanishes. This causes heating, and after final ramping into a detection trap at large distance we have no clear signature of the phase transition to BEC.

Table C.1 summarizes the trap parameters for all phases of the manoeuvre.

Trap	Δt_R [ms]	I_I [A]	I_D	I_S	$B_{b,x}$ [G]	$B_{b,y}$	f_z [kHz]	f_x	$z_{t,0}$ [μm]	x_t
RFC B	600	1.855	0.40	0.0	9.0	37	3.86	0.43	105.9	0
MTrap7	50	1.855	0.40	0.0	18.0	47	2.70	0.62	80.8	0
MTrap8	100	1.500	0.00	-0.1	12.0	40.0	1.80	0.03	79.1	-251
MTrap9	60	1.100	0.00	-0.1	20	44	2.19	0.02	51.5	-305
RFCoolingC	60	1.100	0.05	0.0	11	44	3.36	0.42	51.3	0
MTrap11	3	1.855	0.40	0.0	19.8	60.6	4.81	0.87	63.1	0
MTrap12	3	1.100	0.05	0.0	11	44	3.36	0.42	51.3	0
MTrap13	80	1.100	0.00	0.15	20	44	2.88	0.74	49.5	90
Detection	40	0.400	0.00	0.1	4	7	0.72	0.18	115	93

Table C.1.: Magnetic traps for the transport around the cantilever. Given are the duration of the phase Δt, Ioffe wire current I_I, Dimple wire current I_D, shift wire current I_S, axial (f_x) and transversal ($f_z \approx f_y$) trap frequency, trap-wire distance $z_{t,0}$, and axial trap position x_t.

D. Chip Fabrication

Electroplating process

Material used: Aluminium Nitride AlN, 800 μm thick, one side polished to $R_a < 40$ nm, purchased from Ceramtek or Anceram. Thermal conductivity 180 W/m K.

1. Substrate preparation
   a) raw cleaving
      - Score outline with sharp diamond scorer and cleave with a bench vice.
      - maximum substrate size for UHV evaporation chamber: 52 mm × 43 mm.
      - if possible, 5 mm rim on all sides for handling, mounting, contacting.
      - bevel edges of substrate on wet diamond whetstone.
   b) clean substrate
      - dip for 5 min in Acetone in ultrasonic bath (USB) at 55°C.
      - without drying dip or rinse with Isopropanol, blow dry with clean Nitrogen.
      - dip for 5 min in Piranha (H_2SO_4:H_2O_2 4:1).
      - rinse well in H_2O.
      - dip in Acetone, rinse with Isopropanol, blow dry.
2. Seed layer fabrication
   a) Evaporate gold seed layer.
      - clean $5-10$ min in oxygen plasma ($55-60$ W, 2 Torr O_2), blow dust off.
      - mount in UHV vacuum chamber without delay.
      - turn on e-beam with current set to 55 mA and shutter closed for a few minutes, until pressure drops markedly, typically to $4-10 \times 10^{-9}$ mbar.
      - evaporate 2 nm Ti with evaporation rate $0.1-0.5$ Å/s.

- heat gold with e-beam with shutter closed, current 60 mA.
- evaporate 30 − 50 nm gold at a rate < 1.6 Å/s, optimally 1 Å/s.

b) spin on photo resist
- heat chip 5 min on hotplate at 105°C to desorb water.
- Dispense photo resist ma-P 1240 (micro resist) or ma-P 240 on chip, completely cover surface, avoid bubbles.
- spin on 5 s with 800 rpm, then 40 s with 1000 − 6000 rpm (depending on desired resist thickness, max. ∼ 8 μm possible).
- 5 min softbake at 102°C (depends on resist thickness).

c) Resist exposure and development
- place Cr mask in Mask Aligner, place chip in Mask Aligner on a rubber ring, Newton rings should be visible when contacted.
- Expose for 96 s for Cr mask, 300 s for foil mask.
- Prepare developer (ma-D 336 : H_2O 3:7).
- Rinse chip in water, then dip in developer for 2 : 30 − 3 : 30 min for Cr mask, 5 − 6 min for foil mask, agitate during developing.
- Stop development in beaker with water, rinse ∼ 1 min in running water to remove developer completely.

3. Electroplating

a) Heat electroplating solution (METAKEM ammonium gold sulfide solution $(NH_4)_3Au(SO_3)_2$) in water bath to 57 ± 1°C. Use magnetic stir bar at 100 rpm to agitate solution. Rinse empty bottles with water.

b) Wipe off photoresist with Acetone in chip corners to get contact pads for electric contacting during electroplating.

c) Remove residual photoresist in developed areas by oxygen plasma (50 W, 2 Torr, 5 min).

d) Mount chip on chip holder without delay, connect contact pads, rinse with water.

e) Connect chip to −-pole, platin electrode to +-pole of current source.

f) Set output current such that current density ≤ 3 mA/cm^2, for our chips $I \sim 10 - 40$ mA. The volume growth rate is 10^{-10} m^3/As. Make shure that no uncontrolled voltages or currents occur during switching.

g) Electroplate until desired layer thickness is reached. E.g. for area 1400 mm^2 and 5 μm layer thickness take $I = 34$ mA, $t = 2060$ s.

h) Rinse chip thoroughly with running water, filter electroplating solution with paper filter, refill bottle with water; rinse all equipment thorougly with running water to avoid drying of solution.

4. Remove photoresist, etch seed layer and finish chip
    a) Remove resist in Acetone USB (5 min, 10%, 55°C), rinse in Isopropanol, blow dry.
    b) remove residual resist with Piranha (4:1, 1 min), rinse in water, blow dry.
    c) Etch gold seed layer and Ti adhesion layer with fresh aqua regia (H_2O : H_2Cl : HNO_3 = 1:3:1) for 40 − 120 s (about twice the time until the substrate becomes visible), rinse with water, blow dry.
    d) Measure gold film thickness under microscope.
    e) Spin on photoresist for protection.
    f) Cleave chip to final size.
    g) Bevel chip corners. Marked bevelling E.g. for experiment chip helps to absorb excess glue and avoids glue rims.
    h) Remove resist in Acetone USB (5 min, 10%, 55°C), rinse in Isopropanol, blow dry.

Bibliography

[1] S. Chu, *Cold atoms and quantum control*, Nature **416**, 206 (2002).

[2] A. L. Migdall, J. V. Prodan, and W. D. Phillips, *First Observation of Magnetically Trapped Neutral Atoms*, Phys. Rev. Lett. **54**, 2596 (1985).

[3] R. Grimm, M. Weidemüller, and Y. B. Ovchinnikov, *Optical dipole traps for neutral atoms*, Adv. At. Mol. Opt. Phys. **42**, 95 (2000).

[4] M. H. Anderson, J. R. Ensher, M. R. Matthews, C. E. Wieman, and E. A. Cornell, *Observation of Bose-Einstein Condensation in a Dilute Atomic Vapor*, Science **269**, 198 (1995).

[5] K. B. Davis, M.-O. Mewes, M. R. Andrews, N. J. van Druten, D. S. Durfee, D. M. Kurn, and W. Ketterle, *Bose-Einstein Condensation in a Gas of Sodium Atoms*, Phys. Rev. Lett. **75**, 3969 (1995).

[6] W. Ketterle, D. S. Durfee, and D. M. Stamper-Kurn, *Making, probing and understanding Bose-Einstein condensates*, in *Bose-Einstein condensation in atomic gases, Proceedings of the International School of Physics "Enrico Fermi", Course CXL*, edited by M. Inguscio, S. Stringari, and C. E. Wieman, pages 67–176, Amsterdam, 1999, IOS Press.

[7] Y.-J. Wang, D. Z. Anderson, V. M. Bright, E. A. Cornell, Q. Diot, T. Kishimoto, M. Prentiss, R. A. Saravanan, S. R. Segal, and S. Wu, *Atom Michelson Interferometer on a Chip Using a Bose-Einstein Condensate*, Phys. Rev. Lett. **94**, 090405 (2005).

[8] S. Hofferberth, I. Lesanovsky, B. Fischer, J. Verdu, and J. Schmiedmayer, *Radiofrequency-dressed-state potentials for neutral atoms*, Nat. Phys. **2**, 710 (2006).

[9] P. Böhi, M. Riedel, J. Hoffrogge, J. Reichel, T. W. Hänsch, and P. Treutlein, *Coherent manipulation of Bose–Einstein condensates with state-dependent microwave potentials on an atom chip*, Nat. Phys. **5**, 592 (2009).

[10] D. M. Harber, H. J. Lewandowski, J. M. McGuirk, and E. A. Cornell, *Effect of cold collisions on spin coherence and resonance shifts in a magnetically trapped ultracold gas*, Phys. Rev. A **66**, 053616 (2002).

[11] P. Treutlein, P. Hommelhoff, T. Steinmetz, T. W. Hänsch, and J. Reichel, *Coherence In Microchip Traps*, Phys. Rev. Lett. **92**, 203005 (2004).

[12] B. Julsgaard, A. Kozhekin, and E. S. Polzik, *Experimental long-lived entanglement of two macroscopic objects*, Nature **413**, 400 (2001).

[13] J. Estève, C. Gross, A. Weller, S. Giovanazzi, and M. K. Oberthaler, *Squeezing and Entanglement in a Bose-Einstein condensate*, Nature **455**, 1216 (2008).

[14] M. H. Schleier Smith, I. D. Leroux, and V. Vuletić, *States of an Ensemble of Two-Level Atoms with Reduced Quantum Uncertainty*, Phys. Rev. Lett. **104**, 073604 (2010).

[15] M. F. Riedel, P. Böhi, Y. Li, T. W. Hänsch, A. Sinatra, and P. Treutlein, *Atom chip based generation of entanglement for quantum metrology*, Nature **464**, 1170 (2010).

[16] J. Reichel, W. Hänsel, P. Hommelhoff, and T. W. Hänsch, *Applications of integrated magnetic microtraps*, Appl. Phys. B **72**, 81 (2001).

[17] R. Folman, P. Krüger, J. Schmiedmayer, J. Denschlag, and C. Henkel, *Microscopic Atom Optics: From Wires to an Atom Chip*, Adv. At. Mol. Opt. Phys. **48**, 263 (2002).

[18] J. Reichel, *Microchip traps and Bose-Einstein condensation*, Appl. Phys. B **74**, 469 (2002).

[19] J. Fortágh and C. Zimmermann, *Magnetic microtraps for ultracold atoms*, Rev. Mod. Phys. **79**, 235 (2007).

[20] H. J. Lewandowski, D. M. Harber, D. L. Whitaker, and E. A. Cornell, *Simplified System for Creating a Bose–Einstein Condensate*, J. Low Temp. Phys. **132**, 309 (2003).

[21] E. W. Streed, A. P. Chikkatur, T. L. Gustavson, M. Boyd, Y. Torii, D. Schneble, G. K. Campbell, D. E. Pritchard, and W. Ketterle, *Large atom number Bose-Einstein condensate machines*, Rev. Sci. Instr. **77**, 023106 (2006).

[22] S. Wildermuth, S. Hofferberth, I. Lesanovsky, E. Haller, L. M. Andersson, S. Groth, I. Bar-Joseph, P. Krüger, and J. Schmiedmayer, *Microscopic magnetic-field imaging*, Nature **435**, 440 (2005).

[23] S. Aigner, L. Della Pietra, Y. Japha, O. Entin-Wohlman, T. David, R. Salem, R. Folman, and J. Schmiedmayer, *Long-Range Order in Electronic Transport through Disordered Metal Films*, Science **319**, 1226 (2008).

[24] Y. Lin, I. Teper, C. Chin, and V. Vuletić, *Impact of the Casimir-Polder Potential and Johnson Noise on Bose-Einstein Condensate Stability Near Surfaces*, Phys. Rev. Lett. **92**, 050404 (2004).

[25] J. M. McGuirk, D. M. Harber, J. M. Obrecht, and E. A. Cornell, *Alkali-metal adsorbate polarization on conducting and insulating surfaces probed with Bose-Einstein condensates*, Phys. Rev. A **69**, 062905 (2004).

[26] D. M. Harber, J. M. Obrecht, J. M. McGuirk, and E. A. Cornell, *Measurement of the Casimir-Polder force through center-of-mass oscillations of a Bose-Einstein condensate*, Phys. Rev. A **72**, 033610 (2005).

[27] J. M. Obrecht, R. J. Wild, M. Antezza, L. P. Pitaevskii, S. Stringari, and E. A. Cornell, *Measurement of the Temperature Dependence of the Casimir-Polder Force*, Phys. Rev. Lett. **98**, 063201 (2007).

[28] M. P. A. Jones, C. J. Vale, D. Sahagun, B. V. Hall, and E. A. Hinds, *Spin coupling between cold atoms and the thermal fluctuations of a metal surface*, Phys. Rev. Lett. **91**, 080401 (2003).

[29] D. M. Harber, J. M. McGuirk, J. M. Obrecht, and E. A. Cornell, *Thermally Induced Losses in Ultra-Cold Atoms Magnetically Trapped Near Room-Temperature Surfaces*, J. Low Temp. Phys. **133**, 229 (2003).

[30] H. J. Kimble, *Strong Interactions of Single Atoms and Photons in Cavity QED*, Physica Scripta **176**, 127 (1996).

[31] P. Rabl, D. DeMille, J. M. Doyle, M. D. Lukin, R. J. Schoelkopf, and P. Zoller, *Hybrid Quantum Processors: Molecular Ensembles as Quantum Memory for Solid State Circuits*, Phys. Rev. Lett. **97**, 033003 (2006).

[32] J. Verdú, H. Zoubi, C. Koller, J. Majer, H. Ritsch, and J. Schmiedmayer, *Strong Magnetic Coupling of an Ultracold Gas to a Superconducting Waveguide Cavity*, Phys. Rev. Lett. **103**, 043603 (2009).

[33] A. S. Sørensen, C. H. van der Wal, L. I. Childress, and M. D. Lukin, *Capacitive Coupling of Atomic Systems to Mesoscopic Conductors*, Phys. Rev. Lett. **92**, 063601 (2004).

[34] L. Tian, P. Rabl, R. Blatt, and P. Zoller, *Interfacing Quantum-Optical and Solid-State Qubits*, Phys. Rev. Lett. **92**, 247902 (2004).

[35] D. J. Wineland, C. Monroe, W. M. Itano, D. Leibfried, B. E. King, and D. M. Meekhof, *Experimental Issues in Coherent Quantum-State Manipulation of Trapped Atomic Ions*, J. Res. Natl. Inst. Stand. Technol. **103**, 259 (1998).

[36] L. Tian and P. Zoller, *Coupled Ion-Nanomechanical Systems*, Phys. Rev. Lett. **93**, 266403 (2004).

[37] W. K. Hensinger, D. W. Utami, H.-S. Goan, K. Schwab, C. Monroe, and G. J. Milburn, *Ion trap transducers for quantum electromechanical oscillators*, Phys. Rev. A **72**, 041405(R) (2005).

[38] D. Meiser and P. Meystre, *Coupled dynamics of atoms and radiation-pressure-driven interferometers*, Phys. Rev. A **73**, 033417 (2006).

[39] P. Treutlein, D. Hunger, S. Camerer, T. W. Hänsch, and J. Reichel, *Bose-Einstein Condensate Coupled to a Nanomechanical Resonator on an Atom Chip*, Phys. Rev. Lett. **99**, 140403 (2007).

[40] S. Singh, M. Bhattacharya, O. Dutta, and P. Meystre, *Coupling nanomechanical cantilevers to dipolar molecules*, Phys. Rev. Lett. **101**, 263603 (2008).

[41] C. Genes, D. Vitali, and P. Tombesi, *Emergence of atom-light-mirror entanglement inside an optical cavity*, Phys. Rev. A **77**, 050307 (2008).

[42] H. Ian, Z. R. Gong, Y. Liu, C. P. Sun, and F. Nori, *Cavity optomechanical coupling assisted by an atomic gas*, Phys. Rev. A **78**, 013824 (2008).

[43] A. A. Geraci and J. Kitching, *Ultracold mechanical resonators coupled to atoms in an optical lattice*, Phys. Rev. A **80**, 032317 (2009).

[44] A. B. Bhattacherjee, *Cavity quantum optomechanics of ultracold atoms in an optical lattice: Normal-mode splitting*, Phys. Rev. A **80**, 043607 (2009).

[45] K. Hammerer, M. Aspelmeyer, E. S. Polzik, and P. Zoller, *Establishing Einstein-Podolsky-Rosen Channels between Nanomechanics and Atomic Ensembles*, Phys. Rev. Lett. **102**, 020501 (2009).

[46] K. Hammerer, M. Wallquist, C. Genes, M. Ludwig, F. Marquardt, P. Treutlein, P. Zoller, J. Ye, and H. J. Kimble, *Strong Coupling of a Mechanical Oscillator and a Single Atom*, Phys. Rev. Lett. **103**, 063005 (2009).

[47] K. Hammerer, K. Stannigel, C. Genes, M. Wallquist, P. Zoller, P. Treutlein, S. Camerer, D. Hunger, and T. W. Hänsch, *Optical Lattices with Micromechanical Mirrors*, preprint arxiv:1002.4646 [quant-ph] (2010).

[48] K. Zhang, W. Chen, M. Bhattacharya, and P. Meystre, *Hamiltonian chaos in a coupled BEC-optomechanical-cavity system*, Phys. Rev. A **81**, 013802 (2010).

[49] A. N. Cleland, *Nanoelectromechanical Systems*, Science **290**, 1532 (2000).

[50] A. N. Cleland, *Foundations of Nanomechanics*, Springer-Verlag, Berlin, Germany, 2003.

[51] K. L. Ekinci and M. L. Roukes, *Nanoelectromechanical systems*, Rev. Sci. Instrum. **76**, 061101 (2005).

[52] D. F. Wang, T. Ono, and M. Esashi, *Thermal treatments and gas adsorption influences on nanomechanics of ultra-thin silicon resonators for ultimate sensing*, Nanotechnology **15**, 1851 (2004).

[53] B. M. Zwickl, W. E. Shanks, A. M. Jayich, C. Yang, A. C. B. Jayich, J. D. Thompson, and J. G. E. Harris, *High quality mechanical and optical properties of commercial silicon nitride membranes*, Appl. Phys. Lett. **92**, 103125 (2008).

Bibliography

[54] S. S. Verbridge, H. G. Craighead, and J. M. Parpia, *A megahertz nanomechanical resonator with a room temperature quality factor over a million*, Appl. Phys. Lett. **92**, 013112 (2008).

[55] D. J. Wilson, C. A. Regal, S. B. Papp, and H. J. Kimble, *Cavity Optomechanics with Stoichiometric SiN Films*, Phys. Rev. Lett. **103**, 207204 (2009).

[56] O. Arcizet, P.-F. Cohadon, T. Briant, M. Pinard, and A. Heidmann, *Radiation-pressure cooling and optomechanical instability of a micro-mirror*, Nature **444**, 71 (2006).

[57] G. Anetsberger, O. Arcizet, Q. P. Unterreithmeier, R. Rivière, A. Schliesser, E. M. Weig, J. P. Kotthaus, and T. J. Kippenberg, *Near-field optomechanics with nanomechanical oscillators*, Nature Physics **5**, 909 (2009).

[58] J. D. Teufel, T. Donner, M. A. Castellanos-Beltran, J. W. Harlow, and K. W. Lehnert, *Nanomechanical motion measured with an imprecision below that at the standard quantum limit*, Nature Nanotech. **4**, 820 (2009).

[59] C. Höhberger Metzger and K. Karrai, *Cavity cooling of a microlever*, Nature **432**, 1002 (2004).

[60] A. Schliesser, R. Riviére, G. Anetsberger, O. Arcizet, and T. J. Kippenberg, *Resolved-sideband cooling of a micromechanical oscillator*, Nature Physics **4**, 939 (2008).

[61] Q. P. Unterreithmeier, T. Faust, and J. P. Kotthaus, *Nonlinear Switching Dynamics in a Nanomechanical Resonator*, Phys. Rev. B **81** (2010).

[62] D. Rugar, R. Budakian, H. J. Mamin, and B. W. Chui, *Single Spin Detection by magnetic resonance force microscopy*, Nature **430**, 329 (2004).

[63] C. L. Degen, M. Poggio, H. J. Mamin, C. T. Rettner, and D. Rugar, *Nanoscale magnetic resonance imaging*, PNAS **106**, 1313 (2009).

[64] K. Jensen, K. Kim, and A. Zettl, *An atomic-resolution nanomechanical mass sensor*, Nature Nanotech. **3**, 533 (2008).

[65] T. J. Kippenberg and K. Vahala, *Cavity Optomechanics*, Opt. Exp. **15**, 17172 (2007).

[66] T. J. Kippenberg and K. Vahala, *Cavity Optomechanics: Back-Action at the Mesoscale*, Science **321**, 1172 (2008).

[67] F. Marquardt and S. M. Girvin, *Optomechanics*, Physics **2**, 40 (2009).

[68] I. Favero and K. Karrai, *Optomechanics of deformable optical cavities*, Nature Photon. **3**, 201 (2009).

[69] A. Schliesser, O. Arcizet, R. Riviére, G. Anetsberger, and T. J. Kippenberg, *Resolved-sideband cooling and position measurement of a micromechanical oscillator close to the Heisenberg uncertainty limit*, Nature Physics **5**, 509 (2009).

[70] Y.-S. Park and H. Wang, *Resolved-sideband and cryogenic cooling of an optomechanical resonator*, Nature Physics **5**, 489 (2009).

[71] S. Gröblacher, J. B. Hertzberg, M. R. Vanner, G. D. Cole, S. Gigan, K. C. Schwab, and M. Aspelmeyer, *Demonstration of an ultracold micro-optomechanical oscillator in a cryogenic cavity*, Nature Physics **5**, 485 (2009).

[72] B. A. et al (LIGO scientific collaboration), *Observation of a kilogram-scale oscillator near its quantum ground state*, New J. Phys. **11**, 073032 (2009).

[73] Y.-J. Wang, M. Eardley, S. Knappe, J. Moreland, L. Hollberg, and J. Kitching, *Magnetic Resonance in an Atomic Vapor Excited by a Mechanical Resonator*, Phys. Rev. Lett. **97**, 227602 (2006).

[74] T. Steinmetz, Y. Colombe, D. Hunger, T. W. Hänsch, A. Balocchi, R. J. Warburton, and J. Reichel, *Stable fiber-based Fabry-Pérot cavity*, Appl. Phys. Lett. **89**, 111110 (2006).

[75] Y. Colombe, T. Steinmetz, G. Dubois, F. Linke, D. Hunger, and J. Reichel, *Strong atom-field coupling for Bose-Einstein condensates in an optical cavity on a chip*, Nature **450**, 272 (2007).

[76] I. Favero, S. Stapfner, D. Hunger, P. Paulitschke, J. Reichel, H. Lorenz, E. M. Weig, and K. Karrai, *Fluctuating nanomechanical system in a high finesse optical microcavity*, Opt. Express **17**, 12813 (2009).

[77] D. Hunger, T. Steinmetz, Y. Colombe, C. Deutsch, T. W. Hänsch, and J. Reichel, *Fiber Fabry-Perot cavity with high finesse*, to be published in New Journal of Physics, arXiv: 1005.0067 [physics.optics] (2010).

[78] D. Hunger, C. Deutsch, R. Warburton, T. W. Hänsch, and J. Reichel, *CO_2 laser fabrication of concave, low-roughness depressions on optical fiber endfacets*, in preparation (2010).

[79] D. Hunger, S. Camerer, T. W. Hänsch, D. König, J. P. Kotthaus, J. Reichel, and P. Treutlein, *"Resonant Coupling of a Bose-Einstein Condensate to a Micromechanical Oscillator"*, Phys. Rev. Lett. **104**, 143002 (2010).

[80] W. Hänsel, P. Hommelhoff, T. W. Hänsch, and J. Reichel, *Bose-Einstein condensation on a microelectronic chip*, Nature **413**, 498 (2001).

[81] J. Reichel and J. H. Thywissen, *Using magnetic chip traps to study Tonks-Girardeau quantum gases*, J. Phys. IV France **116**, 265 (2004).

[82] P. Treutlein, *Coherent manipulation of ultracold atoms on atom chips*, PhD thesis, Ludwig-Maximilians-Universität München and Max-Planck-Institut für Quantenoptik, 2008, published as MPQ report 321.

[83] W. Hänsel, J. Reichel, P. Hommelhoff, and T. W. Hänsch, *Magnetic Conveyor Belt for Transporting and Merging Trapped Atom Clouds*, Phys. Rev. Lett. **86**, 608 (2001).

Bibliography

[84] P. Hommelhoff, W. Hänsel, T. Steinmetz, T. W. Hänsch, and J. Reichel, *Transporting, splitting and merging of atomic ensembles in a chip trap*, New J. Phys. **7**, 3 (2005).

[85] S. Gov, S. Shtrikman, and H. Thomas, *Magnetic trapping of neutral particles: Classical and quantum-mechanical study of a Ioffe-Pritchard type trap*, J. Appl. Phys. **87**, 3989 (2000).

[86] D. A. Steck, *Rubidium 87 D Line Data*, http://steck.us/alkalidata/ , version 1.6 (2003).

[87] W. Hänsel, *Magnetische Mikrofallen für Rubidiumatome*, PhD thesis, Ludwig-Maximilians-Universität München and Max-Planck-Institut für Quantenoptik, 2001, published as MPQ report 263.

[88] J. D. Weinstein and K. G. Libbrecht, *Microscopic magnetic traps for neutral atoms*, Phys. Rev. A **52**, 4004 (1995).

[89] J. Fortágh, A. Grossmann, C. Zimmermann, and T. W. Hänsch, *Miniaturized Wire Trap for Neutral Atoms*, Phys. Rev. Lett. **81**, 5310 (1998).

[90] S. Gupta, K. L. Moore, K. W. Murch, and D. M. Stamper-Kurn, *Cavity Nonlinear Optics at Low Photon Numbers from Collective Atomic Motion*, Phys. Rev. Lett. **99**, 213601 (2007).

[91] F. Brennecke, S. Ritter, T. Donner, and T. Esslinger, *Cavity Optomechanics with a Bose-Einstein Condensate*, Science **322**, 235 (2008).

[92] K. W. Murch, K. Moore, S. Gupta, and D. M. Stamper-Kurn, *Observation of quantum-measurement backaction with an ultracold atomic gas*, Nat. Phys. **4**, 561 (2008).

[93] A. Einstein, *Quantentheorie des einatomigen idealen Gases*, Sitzungsber. Preuss. Akad. Wiss. **3**, 18 (1924).

[94] C. C. Bradley, C. A. Sackett, J. J. Tollett, and R. G. Hulet, *Evidence of Bose-Einstein Condensation in an Atomic Gas with Attractive Interactions*, Phys. Rev. Lett. **75**, 1687 (1995).

[95] A. J. Dalfovo, S. Giorgini, L. P. Pitaevskii, and S. Stringari, *Theory of Bose-Einstein condensation in trapped gases*, Rev. Mod. Phys. **71**, 463 (1999).

[96] H. Miesner, D. N. Stamper-Kurn, M. R. Andrews, D. S. Durfee, S. Inouye, and W. Ketterle, *Bosonic Stimulation in the Formation of a Bose-Einstein Condensate*, Phys. Rev. A **45**, 4241 (1992).

[97] P. S. Julienne, F. H. Mies, E. Tiesinga, and C. J. Williams, *Collisional Stability of Double Bose Condensates*, Phys. Rev. Lett. **78**, 1880 (1997).

[98] K. M. Mertes, J. W. Merrill, R. Carretero-González, D. J. Frantzeskakis, P. G. Kevrekidis, and D. S. Hall, *Nonequilibrium Dynamics and Superfluid Ring Excitations in Binary Bose-Einstein Condensates*, Phys. Rev. Lett. **99**, 190402 (2007).

[99] A. Muñoz Mateo and V. Delgado, *Extension of the Thomas-Fermi approximation for trapped Bose-Einstein condensates with an arbitrary number of atoms*, Phys. Rev. A **74**, 065602 (2006).

[100] A. Muñoz Mateo and V. Delgado, *Extension of the Thomas-Fermi approximation for trapped Bose-Einstein condensates with an arbitrary number of atoms*, Phys. Rev. A **75**, 063610 (2007).

[101] S. Stringari, *Collective Excitations of a Trapped Bose-Condensed Gas*, Phys. Rev. Lett. **77**, 2360 (1996).

[102] R. Ozeri, N. Katz, J. Steinhauer, and N. Davidson, *Colloquium: Bulk Bogoliubov excitations in a Bose-Einstein condensate*, Rev. Mod. Phys. **77**, 187 (2005).

[103] T. Kimura, H. Saito, and M. Ueda, *A Variational Sum-Rule Approach to Collective Excitations of a Trapped Bose-Einstein Condensate*, J. Phys. Soc. Jpn **68**, 1477 (1999).

[104] D. S. Jin, J. R. Ensher, M. R. Matthews, C. E. Wieman, and C. E. Cornell, *Collective Excitations of a Bose-Einstein Condensate in a Dilute Gas*, Phys. Rev. Lett. **77**, 420 (1996).

[105] M. Mewes, M. R. Andrews, N. J. van Druten, D. M. Stamper Kurn, C. G. Townsend, and W. Ketterle, *Collective Excitations of a Bose-Einstein Condensate in a Magnetic Trap*, Phys. Rev. Lett. **77**, 988 (1996).

[106] R. Onofrio, D. S. Durfee, C. Raman, M. Köhl, C. E. Kuklewicz, and W. Ketterle, *Surface Excitations of a Bose-Einstein Condensate*, Phys. Rev. Lett. **84**, 810 (2000).

[107] S. Ritter, F. Brennecke, K. Baumann, T. Donner, C. Guerlin, and T. Esslinger, *Dynamical coupling between a Bose-Einstein condensate and a cavity optical lattice*, Appl. Phys. B **95**, 213 (2009).

[108] V. V. Goldman, I. F. Silvera, and T. Leggett, *Atomic hydrogen in an inhomogeneous magnetic field: Density profile and Bose-Einstein condensation*, Phys. Rev. B **24**, 2870 (1981).

[109] D. A. W. Hutchinson, E. Zaremba, and A. Griffin, *Finite Temperature Excitations of a Trapped Gas*, Phys. Rev. Lett. **78**, 1842 (1997).

[110] F. Gerbier, J. H. Thywissen, S. Richard, M. Hugbart, and n. A. A. P. Bouyer, *Experimental study of the thermodynamics of an interacting trapped Bose-Einstein condensed gas*, Phys. Rev. A **70**, 013607 (2004).

Bibliography

[111] D. S. Jin, M. R. Matthews, J. R. Ensher, C. E. Wieman, and E. A. Cornell, *Temperature-Dependent Damping and Frequency Shifts in Collective Excitations of a Dilute Bose-Einstein Condensate in a Dilute Gas*, Phys. Rev. Lett. **78**, 764 (1997).

[112] D. M. Stamper-Kurn, H.-J. Miesner, S. Inouye, M. R. Andrews, and W. Ketterle, *Collisionless and Hydrodynamic Excitations of a Bose-Einstein condensate*, Phys. Rev. Lett. **81**, 500 (1998).

[113] W. Kohn, *Cyclotron Resonance and de Haas-van Alphen Effect of an interacting electron gas*, Physical Review **123**, 1242 (1961).

[114] Y. Castin and R. Dum, *Bose-Einstein Condensates in Time Dependent Traps*, Phys. Rev. Lett. **77**, 5315 (1996).

[115] H. Ott, J. Fortágh, S. Kraft, A. Günther, D. Komma, and C. Zimmermann, *Nonlinear Dynamics of a Bose-Einstein Condensate in a Magnetic Waveguide*, Phys. Rev. Lett. **91**, 040402 (2003).

[116] H. Ott, J. Fortágh, and C. Zimmermann, *Dynamics of a Bose-Einstein Condensate in an anharmonic trap*, J. Phys. B: At. Mol. Opt. Phys. **36**, 2817 (2003).

[117] Y. Japha and Y. B. Band, *Motion of a condensate in a shaken and vibrating harmonic trap*, Phys. Rev. Lett. **78**, 4675 (1997).

[118] F. K. Abdullaev, R. M. Galimzyanov, and K. N. Ismatullaev, *Collective excitations of a BEC under anharmonic trap position jittering*, J. Phys. B: At. Mol. Opt. Phys. **41**, 015301 (2008).

[119] M. Antezza, L. P. Pitaevskii, and S. Stringari, *Effect of the Casimir-Polder force on the collective oscillations of a trapped Bose-Einstein condensate*, Phys. Rev. A **70**, 053619 (2004).

[120] H. B. G. Casimir and D. Polder, *The Influence of Retardation on the London-van der Waals Forces*, Phys. Rev. **73**, 360 (1948).

[121] S. M. Barnett, A. Aspect, and P. W. Milonni, *On the quantum nature of the Casimir-Polder interaction*, J. Phys. B: At. Mol. Opt. Phys. **33**, L143 (2000).

[122] M. Bordag, U. Mohideen, and V. M. Mostepanenko, *New developments in the Casimir effect*, Phys. Rep. **353**, 1 (2001).

[123] M. Antezza, L. P. Pitaevskii, and S. Stringari, *New Asymptotic Behavior of the Surface-Atom Force out of Thermal Equilibrium*, Phys. Rev. Lett. **95**, 113202 (2005).

[124] F. Zhou and L. Spruch, *van der Waals and retardation (Casimir) interactions of an electron or atom with multilayered walls*, Phys. Rev. A **52**, 297 (1995).

[125] S. Y. Buhmann, D.-G. Welsch, and T. Kampf, *Ground-state van der Waals forces in planar multilayer magnetodielectrics*, Phys. Rev. A **72**, 032112 (2005).

[126] A. M. Reyes and C. Eberlein, *Casimir-Polder interaction between an atom and a dielectric slab*, Phys. Rev. A **80**, 032901 (2009).

[127] J.-Y. Courtois, J.-M. Courty, and J. C. Mertz, *Internal dynamics of multilevel atoms near a vacuum-dielectric interface*, Phys. Rev. A **53**, 1862 (1996).

[128] C. I. Sukenik, M. G. Boshier, D. Cho, V. Sandoghdar, and E. A. Hinds, *Measurement of the Casimir-Polder Force*, Phys. Rev. Lett. **70**, 560 (1993).

[129] V. Sandoghdar, C. I. Sukenik, S. Haroche, and E. A. Hinds, *Spectroscopy of atoms confined to the single node of a standing wave in a parallel-plate cavity*, Phys. Rev. Lett. **53**, 1919 (1996).

[130] A. Landagrin, J. Courtois, G. Labeyrie, N. Vansteenkiste, C. I. Westbrook, and A. Aspect, *Measurement of the van der Waals Force in an Atomic Mirror*, Phys. Rev. Lett. **77**, 1464 (1997).

[131] H. Bender, P. W. Courteille, C. Marzok, C. Zimmermann, and S. Slama, *Direct Measurement of intermediate-range Casimir-Polder potentials*, preprint: arXiv 0910.3837 **3** (2009).

[132] F. Shimizu, *Specular Reflection of Very Slow Metastable Neon Atoms from a Solid Surface*, Phys. Rev. Lett. **86**, 987 (2001).

[133] V. Druzhinina and M. DeKieviet, *Experimental Observation of Quantum Reflection far from Threshold*, Phys. Rev. Lett. **91**, 193202 (2003).

[134] T. A. Pasquini, Y. Shin, M. Saba, A. Schirotzek, D. E. Pritchard, and W. Ketterle, *Quantum Reflection from a Solid Surface at Normal Incidence*, Phys. Rev. Lett. **93**, 223201 (2004).

[135] A. Mody, M. Haggerty, J. M. Doyle, and E. J. Heller, *No-sticking effect and quantum reflection in ultracold collisions*, Phys. Rev. B **64**, 085418 (2001).

[136] I. Langmuir and K. H. Kingdon, *Thermionic effects caused by alkali vapors in vacuum cells*, Science **12**, 58 (1923).

[137] R. L. Gerlach and T. N. Rhodin, *Binding and Charge Transfer Associated with Alkali Metal Adsorption on Single Crystal Nickel Surfaces*, Surface Science **19**, 403 (1970).

[138] R. D. Diehl and R. McGrath, *Current progress in understanding alkali metal adsorption on metal surfaces*, J.Phys.: Condens. Matter **9**, 951 (1997).

[139] K. Horn and M. Scheffler, editors, *Theory of Adsorption on Metal Substrates: Electronic Structure*, volume 2, North Holland, Amsterdam, 2000.

[140] I. Langmuir, *Surface chemistry*, Nobel lecture , 287 (1932).

[141] S. Brunauer, P. H. Emmett, and E. Teller, *Adsorption of Gases in Multimolecular Layers*, J. Am. Chem. Soc. **60**, 309 (1938).

Bibliography

[142] P. Kisliuk, *The Sticking Probability of Gases Chemisorbed On The Surface Of Solids*, J. Phys. Chem. Solids **3**, 95 (1957).

[143] J. B. Camp, T. W. Darling, and R. E. Brown, *Macroscopic variations of surface potentials of conductors*, J. App. Phys. **69**, 7126 (1991).

[144] W. Hänsel, M. Harlander, and R. Blatt, *private communication*, to be submitted (2009).

[145] P. Amore and F. M. Fernández, *Exact and approximate expressions for the period of anharmonic oscillators*, Eur. J. Phys **26**, 589 (2005).

[146] J. Landy and R. Sari, *Centered approach to the period of anharmonic oscillators*, Eur. J. Phys. **28**, 1051 (2007).

[147] W. Ketterle and N. J. van Druten, *Evaporative cooling of atoms*, Adv. At. Mol. Opt. Phys. **37**, 181 (1996).

[148] E. L. Surkov, J. T. M. Walraven, and G. V. Shlyapnikov, *Collisionless motion and evaporative cooling of atoms in magnetic traps*, Phys. Rev. A **53**, 3403 (1996).

[149] D. M. Harber, *Experimental investigation of interactions between ultracold atoms and roomtemperature surfaces*, PhD thesis, University of Colorado, Boulder, 2005.

[150] T. Steinmetz, *Resonator-Quantenelektrodynamik auf einem Mikrofallenchip*, PhD thesis, Ludwig-Maximilians-Universität München and Max-Planck-Institut für Quantenoptik.

[151] R. Fermani, S. Scheel, and P. L. Knight, *Spatial decoherence near metallic surfaces*, Phys. Rev. A **73**, 032902 (2006).

[152] V. Peano, M. Thorwart, A. Kasper, and R. Egger, *Nanoscale atomic waveguides with suspended carbon nanotubes*, Appl. Phys. B **81**, 1075 (2005).

[153] P. Petrov, S. Machluf, S. Younis, R. Macaluso, T. David, B. Hadad, Y. Japha, M. Keil, E. Joselevich, and R. Folman, *Trapping cold atoms using surface-grown carbon nanotubes*, Phys. Rev. A **79**, 043403 (2009).

[154] F. Schwabl, *Quanten Mechanik*, Springer-Verlag, Berlin, Germany, 1988.

[155] H. Friedrich, G. Jacoby, and C. G. Meister, *Quantum reflection by Casimir-van der Waals potential tails*, Phys. Rev. A **65**, 32902 (2002).

[156] T. A. Pasquini, M. Saba, G.-B. Jo, Y. Shin, W. Ketterle, D. E. Pritchard, T. A. Savas, and N. Mulders, *Low Velocity Quantum Reflection of Bose-Einstein Condensates*, Phys. Rev. Lett. **97**, 093201 (2006).

[157] B. Segev, R. Côté, and M. G. Raizen, *Quantum reflection from an atomic mirror*, Phys. Rev. A **56**, R3350 (1997).

[158] R. Côté, B. Segev, and M. G. Raizen, *Retardation effects on quantum reflection from an evanescent-wave atomic mirror*, Phys. Rev. A **58**, 3999 (1998).

[159] C. Monroe, W. Swann, H. Robinson, and C. Wieman, *Very Cold Trapped Atoms in a Vapor Cell*, Phys. Rev. Lett. **65**, 1571 (1990).

[160] E. A. Burt, R. W. Ghrist, C. J. Myatt, M. J. Holland, E. A. Cornell, and C. E. Wieman, *Coherence, Correlations, and Collisions: What One Learns about Bose-Einstein Condensates from Their Decay*, Phys. Rev. Lett. **79**, 337 (1997).

[161] J. Söding, D. Guéry-Odelin, P. Desbiolles, F. Chevy, H. Inamori, and J. Dalibard, *Three-body decay of a rubidium Bose-Einstein condensate*, Appl. Phys. B **69**, 257 (1999).

[162] C. Henkel, S. Pötting, and M. Wilkens, *Loss and heating of particles in small and noisy traps*, Appl. Phys. B **69**, 379 (1999).

[163] C. Henkel, P. Krüger, R. Folman, and J. Schmiedmayer, *Fundamental limits for coherent manipulation on atom chips*, Appl. Phys. B **76**, 173 (2003).

[164] C. Henkel, *Exploring surface interactions with atom chips*, preprint arXiv:quant-ph/0512043 (2005).

[165] S. Scheel, P. K. Rekdal, P. L. Knight, and E. A. Hinds, *Atomic spin decoherence near conducting and superconducting films*, Phys. Rev. A **72**, 042901 (2005).

[166] U. Hohenester, A. Eiguren, S. Scheel, and E. A. Hinds, *Spin-Flip lifetimes in superconducting atom chips: Bardeen-Cooper-Shrieffer versus Eliashberg theory*, Phys. Rev. A **76**, 033618 (2007).

[167] B. Kasch, H. Hattermann, D. Cano, T. E. Judd, S. Scheel, C. Zimmermann, R. Kleiner, D. Kölle, and J. Fortágh, *Lifetime measurements of ultracold clouds near cryogenic surfaces*, preprint arXiv 0906.1369 (2009).

[168] T. A. Savard, K. M. O'Hara, and J. E. Thomas, *Laser-noise-induced heating in far-off resonance optical traps*, Phys. Rev. A **56**, 1095 (1997).

[169] M. E. Gehm, K. M. O'Hara, T. A. Savard, and J. E. Thomas, *Dynamics of noise-induced heating in atom traps*, Phys. Rev. A **58**, 3914 (1998).

[170] J. Hoffrogge, *Mikrowellen-Nahfelder auf Atomchips*, 2007, Diploma Thesis.

[171] G. Binnig, C. F. Quate, and C. Gerber, *Atomic Force Microscope*, Phys. Rev. Lett. **56**, 930 (1986).

[172] L. Gross, F. Mohn, N. Moll, P. Liljeroth, and G. Meyer, *The Chemical Structure of a Molecule Resolved by Atomic Force Microscopy*, Science **325**, 1110 (2009).

[173] H. J. Mamin and D. Rugar, *Sub-attonewton force detection at millikelvin temperatures*, Appl. Phys. Lett. **79**, 3358 (2001).

Bibliography

[174] W. C. Fon, K. C. Schwab, J. M. Worlock, and M. L. Roukes, *Nanoscale, phonon-coupled calorimetry with sub-attojoule/attokelvin resolution*, Nano Lett. **28**, 938 (2005).

[175] A. N. Cleland and M. L. Roukes, *A nanometre-scale mechanical electrometer*, Nature **5**, 1968 (2005).

[176] A. Erbe, C. Weiss, W. Zwerger, and R. H. Blick, *Nanomechanical Resonator Shuttling Single Electrons at Radio Frequencies*, Phys. Rev. Lett. **87**, 096106 (2001).

[177] D. V. Scheible, C. Weiss, J. P. Kotthaus, and R. H. Blick, *Periodic Field Emission from an Isolated Nanoscale Electron Island*, Phys. Rev. Lett. **93**, 186801 (2004).

[178] D. R. Koenig, E. M. Weig, and J. P. Kotthaus, *Ultrasonically driven nanomechanical single-electron shuttle*, Nature Nanotech. **3**, 482 (2008).

[179] A. A. Geraci, S. J. Smullin, D. M. Weld, J. Chiaverini, and A. Kapitulnik, *Improved constraints on non-Newtonian forces at 10 microns*, Phys. Rev. D **78**, 022002 (2008).

[180] J. Yang, T. Ono, and M. Esashi, *Zeptogram-Scale Nanomechanical Mass Sensing*, Sens. Act. **82**, 102 (2000).

[181] K. L. Ekinci, Y. T. Yang, and M. L. Roukes, *Ultimate limits to inertial mass sensing based upon nanoelectromechanical systems*, J. Appl. Phys. **95**, 2682 (2004).

[182] Y. T. Yang, C. Callegari, X. L. Feng, K. L. Ekinci, and M. Roukes, *Zeptogram-Scale Nanomechanical Mass Sensing*, Nano Lett. **6**, 583 (2006).

[183] A. K. Naik, M. S. Hanay, W. K. Hiebert, X. L. Feng, and M. L. Roukes, *Towards single-molecule nanomechanical mass spectrometry*, Nature Nanotech. **4**, 445 (2009).

[184] H. W. C. Postma, I. Kozinsky, A. Husain, and M. L. Roukes, *Dynamic range of nanotube- and nanowire-based electromechanical systems*, Appl. Phys. Lett. **86**, 223105 (2005).

[185] J. S. Aldridge and A. N. Cleland, *Noise-Enabled Precision Measurement of a Duffing Nanomechanical Resonator*, Phys. Rev. Lett. **94**, 156403 (2005).

[186] I. Kozinsky, H. W. C. Postma, O. Kogan, A. Husain, and M. L. Roukes, *Basins of Attraction of a Nonlinear Nanomechanical Resonator*, Phys. Rev. Lett. **99**, 207201 (2007).

[187] Q. P. Unterreithmeier, E. M. Weig, and J. P. Kotthaus, *Universal transduction scheme for nanomechanical systems based on dielectric forces*, Nature **458**, 1001 (2009).

[188] R. Lifshitz and M. L. Roukes, *Thermoelastic damping in micro- and nanomechanical systems*, Phys. Rev. B **61**, 5600 (2000).

[189] O. Arcizet, R. Rivière, A. Schliesser, G. Anetsberger, and T. J. Kippenberg, *Cryogenic properties of optomechanical silica microcavities*, Phys. Rev. A **80**, 021803(R) (2009).

[190] A. N. Cleland and M. L. Roukes, *Noise processes in nanomechanical resonators*, J. App. Phys **92**, 2758 (2002).

[191] M. J. Madou, *Fundamentals of Microfabrication: The Science of Miniaturization*, CRC Press, New York, 2nd edition, 2002.

[192] K. C. Schwab and M. L. Roukes, *Putting Mechanics into Quantum Mechanics*, Phys. Today **58**, 36 (2005).

[193] D. F. McGuigan, C. C. Lam, R. Q. Gram, A. W. Hoffmann, D. H. Douglass, and H. W. Gutche, *Measurements of the mechanical Q of single-crystal silicon at low temperatures*, J. Low Temp. Phys. **30**, 621 (1978).

[194] S. S. Verbridge, R. Ilic, H. G. Craighead, and J. M. Parpia, *Size and frequency dependent gas damping of nanomechanical resonators*, Appl. Phys. Lett. **93**, 013101 (2008).

[195] M. C. Cross and R. Lifshitz, *Elastic wave transmission at an abrupt junction in a thin plate with application to heat transport and vibrations in mesoscopic systems*, Phys. Rev. B **64**, 085324 (2001).

[196] I. Wilson-Rae, *Intrinsic dissipation in nanomechanical resonators due to phonon tunneling*, Phys. Rev. B **77**, 245418 (2008).

[197] G. Anetsberger, R. Rivière, A. Schliesser, O. Arcizet, and T. J. Kippenberg, *Ultralow-dissipation optomechanical resonators on a chip*, Nature Photonics **2**, 627 (2008).

[198] J. Yang, T. Ono, and M. Esashi, *Surface Effects and high quality factors in ultrathin single-crystal silicon cantilevers*, Appl. Phys. Lett. **77**, 3860 (2000).

[199] M. Poggio, C. L. Degen, H. J. Mamin, and D. Rugar, *Feedback Cooling of a Cantilever's Fundamental Mode below 5 mK*, Phys. Rev. Lett. **99**, 017201 (2007).

[200] J. D. Thompson, B. M. Zwickl, A. M. Jayich, F. Marquardt, S. M. Girvin, and J. G. E. Harris, *Strong dispersive coupling of a high-finesse cavity to a micromechanical membrane*, **452**, 72 (2008).

[201] S. Perisanu, P. Vincent, A. Ayari, M. Choueib, S. T. Purcell, M. Bechelany, and D. Cornu, *High Q factor for mechanical resonances of batch-fabricated SiC nanowires*, Appl. Phys. B **90**, 043113 (2007).

[202] A. K. Hüttel, G. A. Steele, B. Witkamp, M. Poot, L. P. Kouwenhoven, and H. S. J. van der Zant, *Carbon Nanotubes as Ultrahigh Quality Factor Mechanical Resonators*, Nano Lett. **9**, 2547 (2009).

Bibliography

[203] C. Chen, S. Rosenblatt, K. I. Bolotin, W. Kalb, P. Kim, I. Kymissis, H. L. Stormer, T. F. Heinz, and J. Hone, *Performance of monolayer graphene nanomechanical resonators with electrical readout*, Nature Nanotech. **4**, 861 (2009).

[204] P. W. Anderson, B. I. Halperin, and C. M. Varma, *Anomalous Low-temperature Thermal Properties of Glasses and Spin-Glasses*, Philosophical Magazine **25**, 1 (1972).

[205] R. O. Pohl, X. Liu, and E. Thompson, *Low-Temperature thermal conductivity and acoustic attenuation in amorphous solids*, Rev. Mod. Phys **74**, 991 (2002).

[206] D. R. Southworth, R. A. Barton, S. S. Verbridge, B. Ilic, A. D. Fefferman, H. G. Craighead, and J. M. Parpia, *Stress and Silicon Nitride: A Crack in the Universal Dissipation of Glasses*, Phys. Rev. Lett. **102**, 225503 (2009).

[207] T. Rocheleau, T. Ndukum, C. Macklin, J. B. Hertzberg, A. A. Clerk, and K. C. Schwab, *Preparation and detection of a mechanical resonator near the ground state of motion*, Nature **463**, 72 (2009).

[208] A. A. Clerk, M. H. Devoret, S. M. Girvin, F. Marquardt, and R. J. Schoelkopf, *Introduction to Quantum Noise, Measurement and Amplification*, Rev. Mod. Phys. **82** (2010).

[209] J. W. Wagner and J. B. Spicer, *Theoretical noise-limited sensitivity of classical interferometry*, J. Opt. Soc. Am. B **4**, 1316 (1987).

[210] R. Loudon, *Quantum Limit on the Michelson Interferometer used for Gravitational-Wave Detection*, Phys. Rev. Lett. **47**, 815 (1981).

[211] K. Yamamoto, D. Friedrich, T. Westphal, S. Goßler, K. Danzmann, R. Schnabel, K. Somiya, and S. L. Danilishin, *Quantum noise of a Michelson-Sagnac interferometer with translucent mechanical oscillator*, preprint: arXiv 0912.2603 [quant-ph] (2009).

[212] C. Metzger, M. Ludwig, C. Neuenhahn, A. Ortlieb, I. Favero, K. Karrai, and F. Marquardt, *Self-induced Oscillations in an Optomechanical System Driven by Bolometric Backaction*, Phys. Rev. Lett. **101**, 133903 (2008).

[213] R. G. Knobel and A. N. Cleland, *Nanometre-scale displacement sensing using a single electron transistor*, Nature **424**, 291 (2003).

[214] M. D. LaHaye, O. Buu, B. Camarota, and K. C. Schwab, *Approaching the Quantum limit of a Nanomechanical Resonator*, Science **304**, 74 (2004).

[215] A. Naik, O. Buu, M. D. LaHaye, A. D. Armour, A. A. Clerk, M. P. Blencowe, and K. C. Schwab, *Cooling a nanomechanical resonator with quantum back-action*, Nature **443**, 193 (2006).

[216] C. A. Regal, J. D. Teufel, and K. W. Lehnert, *Measuring nanomechanical motion with a microwave cavity interferometer*, Nat. Phys. **5**, 555 (2008).

[217] M. Eichenfield, J. Chan, R. M. Camacho, K. J. Vahala, and O. Painter, *Optomechanical crystals*, Nature **462**, 78 (2009).

[218] C. M. Caves, *Quantum-mechanical noise in an interferometer*, Phys. Rev. D **23**, 1693 (1981).

[219] H. Müller-Ebhardt, H. Rehbein, C. Li, Y. Mino, a. S. K. Somiya, K. Danzmann, and Y. Chen, *Quantum-state preparation and macroscopic entanglement in gravitational-wave detectors*, Phys. Rev. A **80**, 043802 (2009).

[220] G. G. Ghirardi, A. Rimini, and T. Weber, *Unified dynamics for microscopic and macroscopic systems*, Phys. Rev. D **34**, 470 (1986).

[221] R. Penrose, in *Mathematical Physics 2000*, International Conference on Mathematical Physics, Imperial College Press, London, 2000.

[222] S. Gröblacher, K. Hammerer, M. R. Vanner, and M. Aspelmeyer, *Observation of strong coupling between a micromechanical resonator and an optical cavity field*, Nature **460**, 724 (2009).

[223] C. Fabre, M. Pinard, S. Bourzeix, A. Heidmann, E. Giacobino, and S. Reynaud, *Quantum-noise reduction using a cavity with a movable mirror*, Phys. Rev. A **49**, 1337 (1994).

[224] K. Jähne, C. Genes, K. Hammerer, M. Wallquist, E. S. Polzik, and P. Zoller, *Cavity-assisted squeezing of a mechanical oscillator*, Phys. Rev. A **79**, 063819 (2009).

[225] M. Pinard, A. Dantan, O. Arcizet, T. Briant, and A. Heidmann, *Entangling movable mirrors in a double-cavity system*, Europhys. Lett. **72**, 747 (2005).

[226] D. Vitali, S. Gigan, A. Ferreira, H. R. Böhm, a. A. G. P. Tombesi, V. Vredal, A. Zeilinger, and M. Aspelmeyer, *Optomechanical Entanglement between a Movable Mirror and a Cavity Field*, Phys. Rev. Lett. **98**, 030405 (2007).

[227] S. Mancini, V. I. Man'ko, and P. Tombesi, *Ponderomotive control of quantum macroscopic coherence*, Phys. Rev. A **55**, 3042 (1997).

[228] A. M. Jayich, J. C. Sankey, B. M. Zwickl, C. Yang, J. D. Thompson, S. M. Girvin, A. A. Clerk, F. Marquardt, and J. G. E. Harris, *Dispersive optomechanics: a membrane inside a cavity*, New J. Phys. **10**, 095008 (2008).

[229] S. Bose, K. Jacobs, and P. L. Knight, *Scheme to probe the decoherence of a macroscopic object*, Phys. Rev. A **59**, 3204 (1999).

[230] W. Marshall, C. Simon, R. Penrose, and D. Bouwmeester, *Towards quantum superpositions of a mechanical resonator*, Phys. Rev. Lett. **91**, 130401 (2003).

[231] A. D. Armour, M. P. Blencowe, and K. C. Schwab, *Entanglement and Decoherence of a Micromechanical Resonator via Coupling to a Cooper-Pair Box*, Phys. Rev. Lett. **88**, 148301 (2002).

[232] I. Bargatin and M. L. Roukes, *Nanomechanical Analog of a Laser: Amplification of Mechanical Oscillations by stimulated Zeeman Transitions*, Phys. Rev. Lett. **91**, 138302 (2003).

[233] A. N. Cleland and M. R. Geller, *Superconducting Qubit Storage and Entanglement with Nanomechanical Resonators*, Phys. Rev. Lett. **93** (2004).

[234] L. Tian, *Entanglement from a nanomechanical resonator weakly coupled to a single Cooper-pair box*, Phys. Rev. B **72**, 195411 (2005).

[235] P. Rabl, P. Cappellaro, M. V. Gurudev Dutt, L. Jiang, J. R. Maze, and M. D. Lukin, *Strong magnetic coupling between an electronic spin qubit and a mechanical resonator*, Phys. Rev. B **79**, 041302(R) (2009).

[236] M. D. LaHaye, J. Suh, P. M. Echternach, K. C. Schwab, and M. L. Roukes, *Nanomechanical measurements of a superconducting qubit*, Nature **459**, 960 (2009).

[237] A. D. O'Connell, M. Hofheinz, M. Ansmann, R. C. Bialczak, M. Lenander, E. Lucero, M. Neeley, D. Sank, H. Wang, M. Weides, J. Wenner, J. M. Martinis, and A. N. Cleland, *Quantum ground state and single-phonon control of a mechanical resonator*, Nature online publication , 08967 (2010).

[238] E. A. Donley, N. R. Claussen, S. T. Thompson, and C. E. Wieman, *Atom-molecule coherence in a Bose-Einstein condensate*, Nature **417**, 529 (2002).

[239] M. Greiner, C. A. Regal, and D. S. Jin, *Bose-Einstein condensate from a Fermi gas*, Nature **426**, 537 (2003).

[240] G. Roati, M. Zaccanti, C. D'Errico, J. Catani, M. Modugno, A. Simoni, M. Inguscio, and G. Modugno, ^{39}K *Bose-Einstein Condensate with Tunable Interactions*, Phys. Rev. Lett. **99**, 010403 (2007).

[241] A. O. Caldeira and A. J. Leggett, *Path Integral Approach to Quantum Brownian Motion*, Physica A **121**, 587 (1983).

[242] W. H. Zurek, *Decoherence and the Transition form Quantum to Classical*, Phys. Today **44**, 36 (1991).

[243] W. H. Zurek, *Decoherence, Einselection, and the quantum origins of the classical*, Rev. Mod. Phys **75**, 716 (2003).

[244] W. Yao, T. A. Knuuttila, K. K. Nummila, J. E. Martikainen, A. S. Oja, and O. V. Lounasmaa, *A Versatile Nuclear Demagnetization Cryostat for Ultralow Temperature Research*, J. Low Temp. Phys. **120**, 121 (2000).

[245] R. van Rooijen, A. Marchenkov, H. Akimoto, O. Andreeva, P. van de Haar, R. Jochemsen, and G. Frossati, *Cryostat for Optical Observations Below 1 mK and in Strong Magnetic Fields*, J. Low Temp. Phys. **124**, 497 (2001).

[246] W. T.Strunz and F. Haake, *Decoherence scenarios from microscopic to macroscopic superpositions*, Phys. Rev. A **67**, 022102 (2003).

[247] R. Penrose, *On Gravity's Role in Quantum State Reduction*, Gen. Relativ. Gravit. **28**, 581 (1996).

[248] W. Dür, C. Simon, and J. I. Cirac, *Effective Size of Certain Macroscopic Quantum Superpositions*, Phys. Rev. Lett. **89**, 210402 (2002).

[249] F. Marquardt, B. Abel, and J. von Delft, *Measuring the size of a quantum superposition of many-body states*, Phys. Rev. A **78**, 012109 (2008).

[250] J. I. Korsbakken, K. B. Whaley, J. Dubois, and I. Cirac, *Measurement-based measure of the size of macroscopic quantum superpositions*, Phys. Rev. A **75**, 042106 (2007).

[251] S. Du, M. B. Squires, Y. Imai, L. Czaia, R. A. Saravanan, V. Bright, J. Reichel, T. W. Hänsch, and D. Z. Anderson, *Atom-chip Bose-Einstein condensation in a portable vacuum cell*, Phys. Rev. A **70**, 053606 (2004).

[252] J. Reichel, W. Hänsel, and T. W. Hänsch, *Atomic Micromanipulation with Magnetic Surface Traps*, Phys. Rev. Lett. **83**, 3398 (1999).

[253] S. Groth, P. Krüger, S. Wildermuth, R. Folman, T. Fernholz, J. Schmiedmayer, D. Mahalu, and I. Bar-Joseph, *Atom chips: Fabrication and thermal properties*, Appl. Phys. Lett. **85**, 2980 (2004).

[254] C. A. J. Putman, B. G. De Grooth, N. F. Van Hulst, and J. Greve, *A detailed analysis of the optical beam deflection technique for use in atomic force microscopy*, J. Appl. Phys. **72**, 6 (1992).

[255] S. Wildermuth, P. Krüger, C. Becker, M. Brajdic, S. Haupt, A. Kasper, R. Folman, and J. Schmiedmayer, *Optimized magneto-optical trap for experiments with ultracold atoms near surfaces*, Phys. Rev. A **69**, 030901 (2004).

[256] L. Ricci, M. Weidemüller, T. Esslinger, A. Hemmerich, C. Zimmermann, V. Vuletic, W. König, and T. W. Hänsch, *A compact grating-stabilized diode laser system for atomic physics*, Opt. Comm. **117**, 541 (1995).

[257] T. W. Hänsch, M. D. Levenson, and A. L. Schawlow, *Complete Hyperfine Structure of a Molecular Iodine Line*, Phys. Rev. Lett. **26**, 946 (1972).

[258] X. Zhu and D. T. Cassidy, *Modulation spectroscopy with a semiconductor diode laser by injection-current modulation*, J. Opt. Soc. Am. B **14**, 1945 (1997).

[259] E. L. Raab, M. Prentiss, A. Cable, S. Chu, and D. E. Pritchard, *Trapping of Neutral Sodium Atoms with Radiation Pressure*, Phys. Rev. Lett. **59**, 2631 (1987).

[260] T. W. Hänsch and A. L. Schawlow, *Cooling of gases by laser radiation*, Opt. Commun. **13**, 68 (1975).

[261] S. Chu, L. Hollberg, J. E. Bjorkholm, A. Cable, and A. Ashkin, *Three-Dimensional Viscous Confinement and Cooling of Atoms by Resonance Radiation Pressure*, Phys. Rev. Lett. **55**, 48 (1985).

[262] C. S. Adams and E. Riis, *Laser cooling and trapping of neutral atoms*, Prog. Quant. Electr. **21**, 1 (1997).

[263] M. O. Scully and M. S. Zubairy, *Quantum Optics*, Cambridge University Press, Cambridge, U.K., 1997.

[264] F. Renzoni, S. Cartaleva, G. Alzetta, and E. Arimondo, *Enhanced absorption Hanle effect in the configuration of crossed laser beam and magnetic field*, Phys. Rev. A **63**, 065401 (2001).

[265] L. D. Landau and E. M. Lifschitz, *Mechanik*, volume 1, Harri Deutsch, Frankfurt am Main, 1997.

[266] A. N. Razvi, X. Z. Chu, R. Alheit, G. Werth, and R. Blümel, *Fractional frequency collective parametric resonances of an ion cloud in a Paul trap*, Phys. Rev. A **58**, R34 (1998).

[267] C. Monroe, E. Cornell, C. Sackett, C. Myatt, and C. Wieman, *Measurement of Cs-Cs Elastic Scattering at T= 30 µK*, Phys. Rev. Lett. **70**, 414 (1993).

[268] C. C. Speake and C. Trenkel, *Forces between Conducting Surfaces due to Spatial Variations of Surface Potential*, Phys. Rev. Lett. **90**, 160403 (2003).

[269] J. M. Obrecht, R. J. Wild, and E. A. Cornell, *Measuring electric fields from surface contaminants with neutral atoms*, Phys. Rev. A **75**, 062903 (2007).

[270] P. Treutlein, *Dispersionsmanagement für Materiewellen*, Diploma Thesis, Universität Konstanz, 2002.

[271] J. Berenger, *A Perfectly Matched Layer for the Absorption of Electromagnetic Waves*, J. Comp. Phys. **114**, 185 (1994).

[272] C. Farrell and U. Leonhardt, *The perfectly matched layer in numerical simulations of nonlinear and matter waves*, J. Opt. B **7**, 1 (2005).

[273] M. M. J. Treacy, T. W. Ebbesen, and J. M. Gibson, *Exceptionally high Young's modulus observed for individual carbon nanotubes*, Phys. Rev. Lett. **98**, 263201 (2007).

[274] P. Poncharal, Z. L. Wang, D. Ugarte, and W. A. de Heer, *Electrostatic Deflections and Electromechanical Resonances of Carbon Nanotubes*, Science **283**, 1513 (1999).

[275] M.-F. Yu, O. Lourie, M. J. Dyer, K. Moloni, T. F. Kelly, and R. S. Ruoff, *Strength and Breaking Mechanism of Multiwalled Carbon Nanotubes Under Tensile Load*, Science **287**, 637 (2000).

[276] B. Babić, J. Furer, S. Sahoo, S. Farhangfar, and C. Schönenberger, *Intrinsic Thermal Vibrations of Suspended Doubly Clamped Single-Wall Carbon Nanotubes*, Nano Lett. **3**, 1577 (2003).

[277] V. Sazonova, Y. Yaish, H. Üstünel, D. Roundy, T. A. Arias, and P. L. McEuen, *A tunable carbon nanotube electromechanical oscillator*, Nature **431**, 284 (2004).

[278] B. Lassagne, Y. Tarakanov, J. Kinaret, D. Garcia-Sanchez, and A. Bachtold, *Coupling Mechanics to Charge Transport in Carbon Nanotube Mechanical Resonators*, Science **325**, 1107 (2009).

[279] G. A. Steele, A. K. Hüttel, B. Witkap, M. Poot, H. B. Meerwaldt, L. P. Kouwenhoven, and H. S. J. van der Zant, *Strong Coupling Between Single-Electron Tunneling and Nanoechanical Motion*, Science **325**, 1103 (2009).

[280] J. S. Bunch, A. M. van der Zande, S. S. Verbridge, I. W. Frank, D. M. Tanenbaum, J. M. Parpia, H. G. Craighead, and P. L. McEuen, *Electromechanical Resonators from Graphene Sheets*, Science **315**, 490 (2007).

[281] R. Fermani, S. Scheel, and P. L. Knight, *Trapping cold atoms near carbon nanotubes: Thermal spin flips and Casimir-Polder potential*, Phys. Rev. A **75**, 062905 (2007).

[282] I. Favero and K. Karrai, *Cavity cooling of a nanomechanical resonator by light scattering*, New J. Phys. **10**, 095006 (2008).

[283] D. I. Schuster, A. A. Houck, J. A. Schreier, A. Wallraff, J. M. Gambetta, A. Blais, L. Frunzio, J. Majer, B. Johnson, M. H. Devoret, S. M. Girvin, and R. J. Schoelkopf, *Resolving photon number states in a superconducting circuit*, Nature **445**, 515 (2007).

[284] T. Nirrengarten, A. Qarry, C. Roux, A. Emmert, G. Nogues, M. Brune, J.-M. Raimond, and S. Haroche, *Realization of a Superconducting Atom Chip*, Phys. Rev. Lett. **97**, 200405 (2006).

[285] T. Mukai, C. Hufnagel, A. Kasper, T. Meno, A. Tsukada, K. Semba, and F. Shimizu, *Persistent Supercurrent Atom Chip*, Phys. Rev. Lett. **98**, 260407 (2007).

[286] D. Kano, B. Kasch, H. Hattermann, R. Kleiner, C. Zimmermann, D. Kölle, and J. Fortágh, *Meissner Effect in Superconducting Atom Microtraps*, Phys. Rev. A **101**, 183006 (2008).

[287] C. Roux, A. Emmert, A. Lupascu, T. Nirrengarten, G. Nogues, M. Brune, J.-M. Raimond, and S. Haroche, *Bose-Einstein condensation on a superconducting atom chip*, Europhys. Lett. **81**, 56004 (2008).

[288] A. Emmert, A. Lupascu, G. Nogues, M. Brune, J.-M. Raimond, and S. Haroche, *Measurement of the trapping lifetime close to a cold metallic surface on a cryogenic atom-chip*, Eur. Phys. J. D **51**, 173 (2009).

[289] F. Shimizu, C. Hufnagel, and T. Mukai, *Stable Neutral Atom Trap with a Thin Superconducting Disc*, Phys. Rev. Lett. **103**, 253002 (2009).

Bibliography

[290] T. Müller, B. Zhang, R. Fermani, K. S. Chan, Z. W. Wang, C. B. Zhang, M. J. Lim, and R. Dumke, *Trapping of ultra-cold atoms with the magnetic field of vortices in a thin film superconducting micro-structure*, preprint: arXiv: 0910.2332 [physics.atom-ph] (2009).

[291] I. Bloch, T. W. Hänsch, and T. Esslinger, *Atom Laser with a cw Output Coupler*, Phys. Rev. Lett. **82**, 3008 (1999).

[292] H. Steck, M. Naraschewski, and H. Wallis, *Output of a Pulsed Atom Laser*, Phys. Rev. Lett. **80**, 1 (1998).

[293] B. W. Shore and P. L. Knight, *The Jaynes-Cummings model*, J. Mod. Opt. **40**, 1195 (1993).

[294] G. Rempe, H. Walther, and N. Klein, *Observation of quantum collapse and revival in a one-atom maser*, Phys. Rev. Lett. **58**, 353 (1987).

[295] M. Brune, E. Hagley, J. Dreyer, X. Maître, A. Maali, C. Wunderlich, J. M. Raimond, and S. Haroche, *Observing the Progressive Decoherence of the "Meter" in a Quantum Measurement*, Phys. Rev. Lett. **77**, 4887 (1996).

[296] R. J. Thompson, G. Rempe, and H. J. Kimble, *Observation of normal-mode splitting for an atom in an optical cavity*, Phys. Rev. Lett. **68**, 1132 (1992).

Danksagung

Diese Arbeit ist keine Einzelleistung, sondern entstanden durch die gemeinsame Vision und Anstrengung eines Teams. Hier möchte ich meinen Dank an alle diejenigen aussprechen, die zum Gelingen dieser Arbeit beigetragen haben.

An erster Stelle möchte ich Herrn Professor Hänsch danken. Insbesondere für die Möglichkeit, in so einer aussergewöhnlichen Gruppe mitarbeiten zu dürfen, und für die Gewährung von so viel Freiraum und wohlwollendem Vertrauen.

Besonderen Dank möchte ich Philipp Treutlein aussprechen, der mir ein herausragender Lehrer, enthusiastischer Motivator, diplomatischer Ratgeber, und sportlicher Herausforderer war.

Ebenfalls ganz besonderer Dank gilt meinem Mitstreiter Stephan Camerer, für die heitere gemeinsame Zeit im Labor, am Berg oder auf dem Rad, den bedeutenden Beitrag zu der hier vorgestellten Arbeit, für die unkonventionell verspielten Ideen, die gedankenschärfenden Diskussionen und das Heranführen an den Großmeister Bach.

Meinen Teamkollegen Pascal Böhi, Max Riedel, Johannes Hoffrogge und Maria Korppi möchte ich danken für die freundliche, lustige Gruppenatmosphäre, die grosse Hilfbereitschaft und für die gemeinsamen Freizeitaktivitäten (wenngleich ich auch selten zur Verfügung war).

Professor Jakob Reichel möchte ich danken für die spannende Zusammenarbeit unter anderem bei den Fasercavities, für die schönen Treffen in Paris und die bereichernden Gespräche von Epoisses über Quanten-Zeno Effekt bis Beethoven.

Ich möchte Professor Jörg Kotthaus danken für die gute Zusammenarbeit, den freizügigen Zugang zu Reinraum Einrichtungen, und nicht zuletzt für die Übernahme des Zweitgutachtens. Dank geht auch an Ivan Favero, Daniel König, Sebastian Stapfner, Stephan Schöffberger, und Bert Lorenz, die mir bei Fragen zur Mikrofabrikation und zur Mikro- und Nanomechanik zur Seite standen.

Die fröhliche und offene Atmosphäre auf unserem Gang hat das Arbeiten immer positiv gestimmt, geprägt durch die (z.T. ehemaligen) Mitglieder der Arbeitsgruppen Hänsch und Weinfurter. Insbesondere danke ich Matthias Taglieber, Arne Voigt, Hannes Brachmann, Louis Costa, Daniel Schenk, Florian Henkel und Michael Krug und für spontane Hilfe und Ausleihdienste, abwechslungsreiche Unterhaltungen und lustige Abende auf Ringberg.

Mit der Arbeitsgruppe von Jakob in Paris verbinden mich nicht nur die spannenden Gruppentreffen, sondern auch die Zusammenarbeit über die Fasercavities und lehrreiche Diskussionen und Tips zu unseren Experimenten. Ich möchte dabei besonders Tilo Steinmetz, Yves Colombe und Christian Deutsch danken.

Toni Scheich danke ich für die Assistenz bei elektronischen Schwierigkeiten, Nicole Schmidt für den bereitwilligen Beistand in organisatorischen, chemischen, oder Softwaretechnischen Fragen, Wolfgang Simon, Charly Linner, Herrn Aust und Herrn Grosshauser sowie dem Team der Metallbearbeitungswerkstätten an der LMU für Design und Fertigung mechanischer Bauteile. Gabriele Gschwendtner danke ich für ihre dynamische Erledigung bürokratischer und organisatorischer Aufgaben sowie für die Kulanz bei meinen regelmässigen Versäumnissen bei der Abgabe terminkritischer Dokumente.

Ohne meine Familie, die mich immer unterstützt und gefördert hat, wäre alles viel schwieriger. Ich danke deswegen ganz besonders Sabine, Jakob, Miriam, meinen Eltern, Geschwistern und Freunden für ihr Zutun.

Die VDM Verlagsservicegesellschaft sucht für wissenschaftliche Verlage abgeschlossene und herausragende

Dissertationen, Habilitationen, Diplomarbeiten, Master Theses, Magisterarbeiten usw.

für die kostenlose Publikation als Fachbuch.

Sie verfügen über eine Arbeit, die hohen inhaltlichen und formalen Ansprüchen genügt, und haben Interesse an einer honorarvergüteten Publikation?

Dann senden Sie bitte erste Informationen über sich und Ihre Arbeit per Email an *info@vdm-vsg.de*.

Sie erhalten kurzfristig unser Feedback!

VDM Verlagsservicegesellschaft mbH
Dudweiler Landstr. 99   Telefon +49 681 3720 174
D - 66123 Saarbrücken   Fax   +49 681 3720 1749
www.vdm-vsg.de

Die VDM Verlagsservicegesellschaft mbH vertritt

Printed by Books on Demand GmbH, Norderstedt / Germany